ELEMENTS
OF
CERAMICS

ADDISON-WESLEY METALLURGY SERIES

Morris Cohen, *Consulting Editor*

ELEMENTS
OF
CERAMICS

F. H. NORTON

Professor of Ceramics
Massachusetts Institute of Technology

1952

ADDISON-WESLEY PRESS, INC.

CAMBRIDGE 42, MASS.

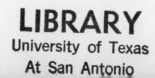

PREFACE

For some time it has been apparent that our undergraduate ceramic courses were in need of a text to cover the broad principles of ceramics based on an approach from crystal physics and unit processes. In this volume an attempt has been made to provide an introduction to ceramics at Junior or Senior level, on top of which may be built more specific engineering courses or more advanced graduate courses.

At the same time it is felt that there are many workers in the industry who would welcome a book such as this to help keep them abreast of the recent developments in the field.

Also, there is the great body of ceramic artists whose thirst for more knowledge is attested to by a constant stream of interviews and correspondence with the author. It is believed that this group will find much of interest in this book, especially the carefully worked out body and glaze formulas and the descriptions of well-tried ceramic kilns.

Throughout this book an effort has been made to carry visual education as far as practicable by the use of carefully prepared illustrations. This effort has been made more effective because the author prepared the illustrations himself along with the text. This method has the twofold advantage of compressing a great deal of information into a small space, and at the same time making it easier for the reader to comprehend.

CONTENTS

INTRODUCTION

Definition

Ceramics may be defined in a somewhat broader sense than the dictionary definition of "pottery" implies. It seems evident that the Greek word *Keramos* meant "burnt stuff"; thus our modern term, ceramics, which includes whitewares, enamels, refractories, glass, cements, fired building materials, and abrasives, is not incompatible with the original usage.

Historical development

Fragments of pottery have been associated with the sites of human dwelling places from very early times; in fact, these sherds have been one of the chief aids to the archeologist in the assignment of cultural levels. The use of fired clay seems to have originated independently in a number of places, rather than to have spread from a single focal point. This does not seem strange, for nothing would be more natural than to daub the ever-present clay onto baskets for cooking over an open fire, or to form this plastic material into such vessels.

There is not space here to discuss the artistic developments in the history of ceramics, but many excellent references covering this subject are available. An attempt has been made, however, to show in the charts of Fig. 1 the technical progress from the earliest times to the present. Since many portions of the picture are not clear, a high degree of precision is not pretended, but these gaps may be filled in by future research.

In looking over these charts one is struck by the great antiquity of some of the objects and processes used today, especially the contributions at the beginning of the Christian Era in Rome and China. One also notes the great development in the 18th century in Europe and, finally, the tremendous strides at the end of the 19th century and the beginning of the 20th as science and engineering were applied to this ancient art.

Scope and magnitude of the industry in the United States

In Fig. 2 is shown a diagram of the ceramics industry as a whole, which gives an idea of its magnitude. Altogether, it is one of the large industries, with an output of about three billion dollars annually. It should be kept in mind that it is closely associated with our daily lives, since many of the essentials in any home are ceramic — the dishes on the table, glass in the windows, insulation in the walls, brick in the chimney, enamel in the bathtub, and porcelain in the electric fixtures. Also, most other industries are dependent on ceramics; for example, the metallurgical industry requires refractories, and the automobile industry needs abrasives.

Literature in the ceramic field

There are many excellent books covering various phases of ceramics, but an attempt is made here to select those of most direct interest to the student at the undergraduate level. At the end of each chapter will be found a list of pertinent references. The author has reluctantly omitted foreign

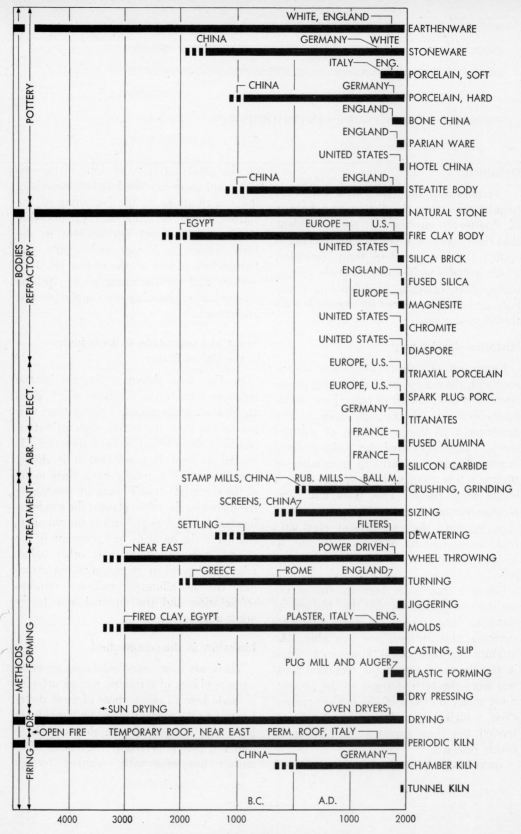

FIG. 1. Technical history of ceramics.

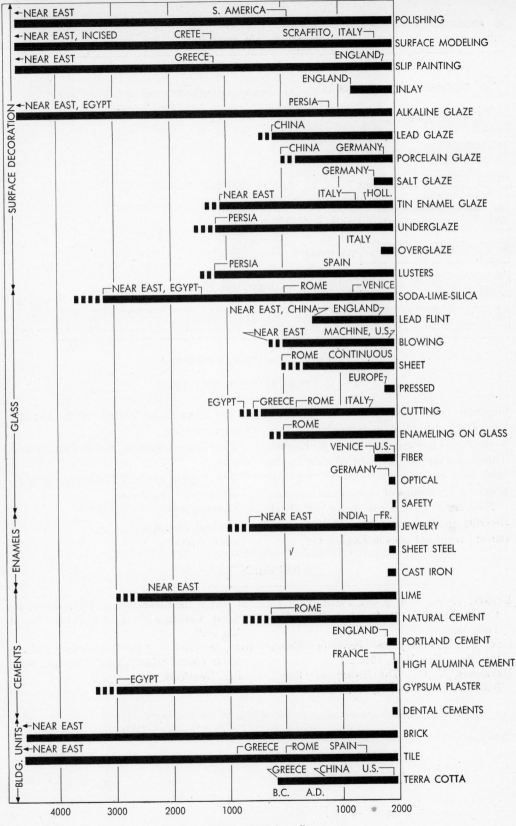

FIG. 1 (continued).

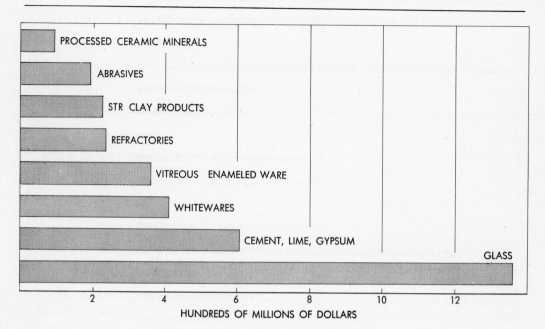

FIG. 2. Scope of the ceramic industry.

language books and periodicals, since so few of our technical undergraduates have a reading knowledge outside of English. This lack narrows the point of view in this field, and ceramic students are urged to acquire a knowledge of French and German.

Periodicals. There are many periodicals covering the ceramic field. The more important technical ones in English are:

The Journal and *Bulletin of the American Ceramic Society*

The Transactions of the British Ceramic Society

The Journal of Glass Technology (Eng.)

Brick and Clay Record

Ceramic Age

Ceramic Industry

Glass Industry

REFERENCES

BARBER, E. A., *Pottery and Porcelain of the United States.* G. P. Putnam's Sons, New York, 1893.

COX, W. E., *Pottery and Porcelain.* Crown Publishers, New York, 1944.

DEVEVOISE, N. C., "The History of Glaze." *Bull. Am. Ceram. Soc.* **13,** 293, 1934.

"Report of the Committee on Definition of the Term 'Ceramics.' " *J. Am. Ceram. Soc.* **3,** 526, 1920.

RIES, H., *History of the Clay-working Industry in the United States.* John Wiley and Sons, Inc., New York, 1909.

CHAPTER 1

CLAY MINERALS

Introduction

Since clay is the backbone of ceramics, it is important to understand the nature of clay itself. It was formerly thought that clay consisted largely of an amorphous material with submicroscopic particles. But as competent men with modern tools examined the finer fractions of clays, it became more and more evident that there was little that might be called amorphous; rather there were present minute, but definite crystals. These crystals could be grouped into fairly definite species of minerals, although perfection of crystallization was found lacking in many cases. We now have a good insight into the minerals making up the clays, although there is still much to be done, especially in the field of hydrous micas.

This chapter attempts to give a brief outline of crystal chemistry, showing how it applies to the silicates in general and to the clay minerals in particular.

Elements of crystal chemistry

Properties of the atoms. The atom may be considered to consist of a nucleus having a net positive charge equal to the atomic number, surrounded by shells of electrons which total up to an equal negative charge. For example, hydrogen has one electron and an atomic number of one; electrons are added, one at a time, to fill shell after shell until uranium, with a total of 92 electrons, is reached. Table 1–1 shows the electrons in each shell for the various atom species,

and it will be seen later that the properties of the element are influenced by the number and location of the electrons.

When atoms or ions (an ion is an atom that has lost or gained one or more electrons) are packed together to form a crystal, each takes up a definite space that may be assumed to be a sphere. The radius of this sphere is known as the ionic radius, values of which are shown in Table 1–2. In a few cases, such as in the lead ions, the electrons are not symmetrically arranged about the nucleus, but are inclined to be lopsided or polarized. This tends to give a different type of packing than that indicated by the value of ionic radius alone, especially on the surface of solids.

The very simple model of the atom de·scribed above is adequate for a preliminary study of crystals and glasses. The real structure of the atom is more complex. There are a number of excellent treatises on the structure of the atom if the reader desires to go deeper into this subject.

Bonding forces between atoms. When atoms are regularly located in the crystal lattice, there must be present forces that hold them in place. These forces are termed *bonds* and generally represent a condition of balance between attraction forces and repulsion forces. Therefore, an additional force is required to move the atoms from their stable separation distance.

One of the important bonding forces in crystals is the ionic bond, which is due to the metallic atoms losing an outer electron

1

Table 1-1

Classification of Atoms according to Structure

Note: Columns 9–17 and 18–31 each carry two outer complete shells of "2 2" electrons.

Complete Shells (K L M N O P)	0	1	2	3	4	5	6	7	9	10	11	12	13	14	15	16	17	18	19	20	21	22	23	24	25	26	27	28	29	30	31
		H																													
2	He	Li	Be	B	C	N	O	F																							
2 8	Ne	Na	Mg	Al	Si	P	S	Cl																							
2 8	A	K	Ca						Sc	Ti	V	Cr*	Mn	Fe	Co	Ni	Cu*														
2 8 18	K		Zn	Ga	Ge	As	Se	Br																							
2 8 18 8		Rb	Sr						Yt	Zr	Cb*	Mo*	Ma*	Ru*	Rh*	Pd*	Ag*														
2 8 18 18	Xe		Cd	In	Sn	Sb	Te	I																							
2 8 18 18 8		Cs	Ba						Lu	Hf	Ta	W	Re	Os	Ir	Pt	Au*														
2 8 18 32																		Lu	Ce	Pr	Nd	Ie	Sm	Eu	Gd	Tb	Dy	Ho	Er	Tu	Yb
2 8 18 32 18	Rn		Hg	Tl	Pb	Bi	Po	-																							
2 8 18 32 18 8		-	Ra																												
2 8 18 21 18									Ac	Th*	Pa	U																			

Band descriptions (Incomplete Shells):

- Column 0: All shells complete — No compounds — No absorption in visible spectrum
- Columns 1–7: One shell incomplete — No absorption in visible spectrum
- Columns 9–17: Two shells incomplete — Variable valence — Broad band absorption in visible spectrum — (Transition elements)
- Columns 18–31: Three incomplete shells — Sharp band absorption in visible spectrum — (Rare earth elements)

*Elements in which the normal atom is believed to have one electron in the outer shell.

Table 1-2

Ionic Radii of the Elements Commonly Used in Ceramics

Element	Atomic number	Ionic radius in Å units	Ion
Aluminum	13	0.57	Al^{+++}
Barium	56	1.43	Ba^{++}
Beryllium	4	0.34	Be^{++}
Calcium	20	1.06	Ca^{++}
Carbon	6	0.20	C^{++++}
Chromium	24	0.64	Cr^{+++}
Cobalt	27	0.82	Co^{++}
Copper	29	0.96	Cu^{+}
Fluorine	9	1.33	F^{-}
Iron	26	0.83	Fe^{++}
Iron	26	0.67	Fe^{+++}
Lead	82	1.32	Pb^{++}
Lithium	3	0.78	Li^{+}
Magnesium	12	0.78	Mg^{++}
Manganese	25	0.91	Mn^{++}
Nickel	28	0.78	Ni^{++}
Oxygen	8	1.32	O^{--}
Phosphorus	15	0.35	P^{+++++}
Potassium	19	1.33	K^{+}
Silicon	14	0.39	Si^{++++}
Sodium	11	0.98	Na^{+}
Strontium	38	1.27	Sr^{++}
Tin	50	0.74	Sn^{++++}
Titanium	22	0.69	Ti^{++++}
Zinc	30	0.83	Zn^{++}
Zirconium	40	0.87	Zr^{++++}

to become positive ions and the nonmetallic atoms gaining an outer electron to form a negative ion. The resulting coulomb attraction holds the atoms together. Simple inorganic salts often have this type of bonding, for example, NaCl, as well as many of the minerals used in ceramics. Ionic bonded crystals are brittle and have medium to high melting points.

Another type of bonding force is the covalent bond, where a pair of electrons is shared by two atoms. Elements such as C, Si, N, P, and O often have the covalent bond. This type of bonding gives hard, strong materials with high melting points.

There is a third type of bond, called the metallic bond, in crystals composed only of positive ions. Here the closely packed atoms are pictured as surrounded by an electron cloud which holds them together. This bond gives more plastic materials with a wide range of melting points.

A fourth type of bond, which can hardly be called chemical bonding, is the van der Waals force. In general, it is weak in character and gives ready cleavage.

In any one crystal it cannot be assumed that the bonding force is exclusively one of those mentioned above; it is more generally a combination of them. This is certainly true in many of the silicates.

Unit cell. A crystal is made up of an orderly arrangement of one or more atomic species, as shown schematically by the two-dimensional network of Fig. 1–1(a). The smallest portion of the crystal network that can be used as a repeating building block to form the whole is called a unit cell, outlined by the dotted lines. The dimensions and the placing of the atoms in the unit cell are the fundamental property of the crystal. Most of the unit cells of the crystals used in ceramics have now been worked out on the basis of the pioneer work of W. L. Bragg and his students. Descriptions of these structures may be found in the *Zeitschrift für Kristallographie* and the references at the end of this chapter.

In contradiction to the uniform pattern of the crystal is the random network of the glassy state as shown in Fig. 1–1(b). This state of solid matter will be discussed in Chapter 17. As an example, the simple unit cell of CsCl type may be illustrated as in Fig. 1–2. Here the cell is cubic with the cations placed at the corners and the anion at the center. As only one-eighth of the cations are included at the corners, the cell represents $8 \times \frac{1}{8}$ cation $+ 1$ anion, which

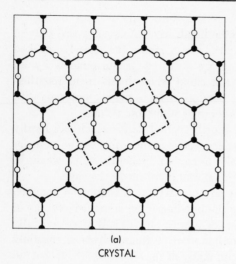

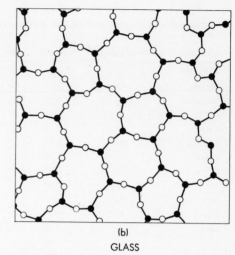

(a)
CRYSTAL

(b)
GLASS

FIG. 1–1. (a) Crystal. (b) Glass.

corresponds with the one to one formula of CsCl. The crystal is built up by repeating this cube over and over.

Anyone studying the atomic arrangement in crystals will be at once struck by the beauty of the patterns. Berkhoff in *Aesthetic Measure* has defined a beautiful design as one where the largest possible proportion

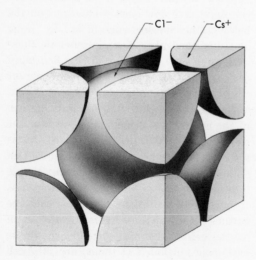

FIG. 1–2. Unit cell of cesium chloride.

of the elements involved show a perceptible sense of order. The crystal network, therefore, is a three-dimensional design of great beauty. In fact, in most crystals with fixed relative numbers of the various atoms, there is only one possible arrangement that will give a stable structure of minimum energy. Each atom must fulfill certain rules in regard to bonding with other atoms. These rules were first set forth by Pauling, but little can be said about them here because of space limitations.

Coordination number. It has been shown that the atoms vary in diameter to a considerable extent. Thus it is possible to place a considerable number of small atoms about a large atom, while only a few large atoms can be placed about a small atom. Therefore, the ratio between the diameter of the cation and that of the anion is an important number, often referred to as the radius ratio. The number of atoms immediately surrounding and bonded to another atom is known as the coordination number. As the radius ratio increases, so does the coordination number, as shown in Table 1–3.

Table 1-3

Relation of Radius Ratio and Coordination Number for
Ions A and X Acting as Rigid Spheres

Radius ratio $R_A : R_X$	Coordination number	Arrangement	Type of structure
	1		Single molecules
			Double molecules
Up to 0.15	2	Opposite each other	Molecular chains
0.15-0.22	3	Corners of an equilateral triangle	Boron nitride
0.22-0.41	4	Corners of a tetrahedron	ZnS
0.41-0.73	6	Corners of an octahedron	NaCl
0.73-1.00	8	Corners of a cube	CsCl
1.00 and above	12	Closest packing	Cu

Thus it can be seen that, for simple structures, where the ions act as rigid spheres, the type of packing in the crystal can be predicted with reasonable certainty. However, in cases where polarization occurs the ions are not spherical and the above rule requires some modification.

Crystal structure. The crystal is built up of a series of unit cells. A crystal starts growing from a nucleus of one or more unit cells forming a tiny group of atoms that may be thought of as the seed. This nucleus, under the proper environmental conditions, will grow by the addition of atoms according to a regular structural pattern. No crystal, however, is so perfect that all the atoms are in their proper places, for here and there are present discontinuities that cause defects in the structure and may have an important influence on the properties of the crystal.

The crystal has an outside form which permits it to be classified into six systems based on symmetry, but not all crystals are so perfect that the system can be determined by visual inspection. However, optical and x-ray studies will usually yield the symmetry relations. The six systems with their axes are shown in Fig. 1-3. There are many excellent treatises on crystallography, so that those interested may easily pursue this fascinating subject further.

Polymorphism. A specific compound such as silica (SiO_2) may occur in several different forms of crystal. In each form the ions and the ratio of the number of cations to anions is the same, but the arrangement differs. Some forms are stable in one temperature range and some in others. Even in small amounts, impurities may influence the occurrence of a particular form. These different crystalline forms of the same material, known as polymorphic forms, are of great interest in ceramics.

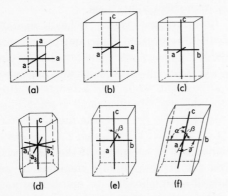

FIG. 1-3. The six crystallographic systems: (a) isometric; (b) tetragonal; (c) orthorhombic; (d) hexagonal; (e) monoclinic; (f) triclinic.

Solid solutions. In many cases it is possible to have a crystal A-X form a continuous series of mixtures with a crystal B-X. This may occur by B ions replacing some of the A ions (substitution solid solution); by A ions, if smaller than B, finding holes in B-X and gradually filling them (interstitial solid solution); or by omission of some ions from the lattice to form holes (omission solid solution). To form a solid solution the two crystals must be similar in geometry. Also, the ionic radius of A must not be more than 15 per cent larger or smaller than that of B. The closer in size they are, the more complete will be the range of solid solution. Also, at higher temperatures, where the crystal lattice is loosened up by thermal agitation, solid solution can occur more readily.

Silicate structures

Up to within the last twenty-five years the structure of the silicates was somewhat of a mystery. However, the work of W. L. Bragg, B. E. Warren, and others clarified the structures by means of x-ray diffraction methods, so that now we have a good picture of the important compounds. The silicates are really the basis on which ceramics is built, and therefore will be emphasized throughout this book.

Silicate types. The fundamental building block in the silicates is the silicon-oxygen tetrahedron with a silicon atom at the center and four oxygens at the corners. These tetrahedrons are fitted together in various ways to form the silicates, as shown in Table 1–4. In the diagrams no attempt is made to indicate the sizes of the ions; only the position of their centers is shown.

Orthosilicates. These structures are of interest as refractories because of their high softening point. The independent tetrahedrons make a structure of good stability.

Disilicates. These sheet structures are of particular interest, since they form the basis of the clay minerals described later in this chapter. This sheet structure causes perfect cleavage parallel to the *ab* plane in the crystal.

Kaolin minerals

Formulas for the clay minerals. In the classic paper on *Minerals of the Montmorillonite Group* by Ross and Hendricks, a logical method of expressing the formula of a clay mineral from its chemical analysis is developed. As this method of writing the formulas will be used in this book, a brief discussion of the reasoning behind it will be necessary.

A general formula for montmorillonite, for example, is

$$\underbrace{[Al^{+3}_{a-y} + Fe^{+3}_{b} + Mg^{+2}_{d}]}_{\substack{\text{Octahedral}\\\text{coordination}}} \quad \underbrace{[Al^{+3}_{y} + Si_{4-y}]}_{\substack{\text{Tetrahedral}\\\text{coordination}}}$$

$$\underbrace{O_{10}[OH_2]}_{\text{Anions}} \quad \underbrace{X_{0.33},}_{\substack{\text{Exchangeable}\\\text{bases}}}$$

where the ions in octahedral positions $= a - y + b + d$, and 0.33 is the amount of the exchangeable bases.

From specific chemical analysis the values of the subscripts may be computed as shown in the paper cited above. For a typical montmorillonite, the expression becomes

$$\left[\begin{array}{c} Na_{0.33} \\ \uparrow \\ Al_{1.67}Mg_{0.33} \end{array}\right] [Si_4 \, O_{10}][OH],$$

which is now completely balanced, the first bracket representing the cations in octahedral coordination.

Referring back to Table 1–2, it will be seen that the ionic radii of the cations and oxygen are

Table 1-4. Structure of the Silicates

	Motif of Si–O tetrahedrons	Si:O ratio of smallest unit	Volume charge of silica unit	Typical mineral	Atomic plan
Orthosilicates	Independent tetrahedrons sharing no oxygens	1:4	-4	Forsterite Mg_2SiO_4	
Pyrosilicates	Independent pairs of tetrahedrons sharing one oxygen	2:7	-6	Akermanite $Ca_2MgSi_2O_7$	
Metasilicate chains	Continuous single chains of tetrahedrons sharing two oxygens	1:3	-2	Diopside $CaMg(SiO_3)_2$	
Metasilicate chains	Continuous double chains of tetrahedrons sharing alternately 2 and 3 tetrahedrons	4:11	-6	Tremolite $H_2Ca_2Mg_5(SiO_3)_8$	
Metasilicate rings	Closed independent rings of tetrahedrons each sharing two oxygens	3.9 6:18	-6 -12	Benitoite $Ba Ti Si_3O_9$ Beryl $Al_2Be_3Si_6O_{18}$	
Disilicates	Continuous sheets of tetrahedrons each sharing two oxygens	4:10	-4	Muscovite $Al_4K_2(Si_6Al_2)O_{20}(OH)_4$	
Silica	Three-dimensional network of tetrahedrons each sharing all four oxygens	1:2 1:2	0 0	Quartz SiO_2 Orthoclase $K Al Si_3O_8$	(Three-dimensional structure not shown)

Ions	Mg^{++}	Al^{+++}	Si^{++++}	O^{--}
Ionic radius, Å's	0.78	0.57	0.39	1.32
$\dfrac{\text{Cation radius}}{\text{O radius}}$	0.60	0.43	0.30	

Looking at Table 1–3, it will be seen that Mg and Al must be in sixfold (octahedral) coordination, while silicon must be in fourfold (tetragonal) coordination.

Kaolinite [$(OH)_4Al_2Si_2O_5$]. The majority of high-grade clays consist largely of the mineral kaolinite. This mineral occurs in tiny flat plates roughly hexagonal in outline, as shown in Fig. 1–4 from an excellent microphotograph taken with an electron microscope. The average size is about 0.7 micron in diameter and about 0.05 micron in thickness. At times these crystals are found in groups known as "books" or

(a)

(b)

FIG. 1–4. A kaolinite crystal 0.7 micron in diameter. The shadow is three times the plate thickness. (Electron microscope photograph from the thesis of Dr. Walter East. Taken by C. E. Hall, Massachusetts Institute of Technology, Cambridge, Massachusetts.)

"worms," where many crystals are piled one on another.

By means of optical and x-ray measurements, it has been possible to arrive at a probable arrangement of the atoms in the structure. As it is difficult to show this structure completely in a two-dimensional diagram, a dissected view is shown in Fig. 1–5.

The crystal is made up of a series of parallel sheets in the plane of the *a, b* axes, built up to the number of fifty in a crystal such as shown in Fig. 1–4. Each sheet is composed of a tetrahedral Si-O layer and

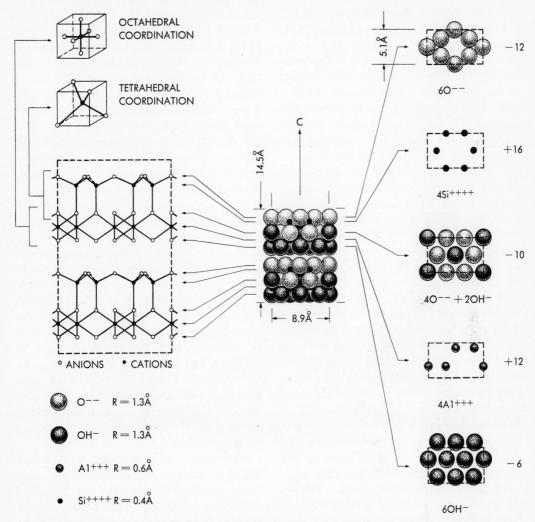

FIG. 1–5. The unit cell of kaolinite, $4[(OH)_4Al_2Si_2O_5]$. In the center is shown a side view of the unit cell enclosed in dotted lines. At the right is a series of horizontal sections of this cell at the heights of the atomic layers. At the left is a schematic diagram of the side of the unit cell magnified two times to show the bonding.

an octahedral Al-OH layer, as shown in Fig. 1–5. While the sheet is unsymmetrical, the total charge adds up to zero, giving a balanced structure. The Si-O layer shows a close packing of the O^{--}, but with holes in the hexagonal structure. On the other hand, the Al-OH layer is open-packed with no holes.

In the diagram, the monoclinic angle is not shown, as it is in a plane normal to the paper, but it actually is 100° because of a shift of one layer over the next. The bonding between layers is due to weak, residual, or van der Waals' forces between the O^{--} and OH^{-} ions, which accounts for the easy cleavage. It can be seen that the unit cell consists of the two sheets containing four molecular weights of $(OH)_4Al_2Si_2O_5$.

If possible, the student should study a three-dimensional model of this mineral, since it is difficult to portray this intricate structure clearly on the printed page. In fact, one of the best ways to become thoroughly familiar with the structure is to actually make models from appropriately sized spheres.

Other kaolin minerals. There are several other minerals that are believed to have a sheet structure nearly identical to that of kaolinite; however, the shift from one sheet to another is in different directions in these minerals. They include dickite and nacrite, very similar to kaolinite, but rather rare constituents of clay, and endellite, of finer grain size, less perfectly crystalline, and perhaps with a different arrangement of the sheets. Finally, there is the more or less amorphous allophane with very little regularity in the atomic arrangement. The latter two minerals are found in relatively pure form in few deposits but may become of some use in ceramics in the future.

Montmorillonite minerals

Montmorillonite

$$\begin{pmatrix} Na_{0.33} \\ \uparrow \\ Al_{1.67}Mg_{0.33} \end{pmatrix} Si_4O_{10}(OH)_2.$$

This mineral is found in bentonite, derived from volcanic ash. It is characterized by very fine platelike particles, seldom over .05 micron in diameter.

The structure of this mineral, as shown in Fig. 1–6, differs from kaolinite in having a symmetrical sheet of an Al-OH octahedral layer sandwiched between two Si-O tetrahedral layers. One of the aluminum ions is replaced by a divalent ion such as magnesium, which gives a net charge for the unit cell of -2. This is balanced by adsorbed ions not shown in the diagram. The sheets are not definitely aligned with one another, so we cannot exactly define a unit cell or the monoclinic angle. The unit cell contains only one lattice sheet and two molecular weights.

Montmorillonite is unique among minerals in that water molecules can force themselves between the sheets and thus cause swelling. They enter at A, Fig. 1–6, since the residual bonds between the O^{--} of one sheet and the O^{--} of the next are very weak as compared with the O^{--}-OH^{-} bonds in kaolinite, and may build up to as many as six molecular layers of H_2O. Also, ions may be adsorbed not only on the edges but also on the faces between sheets, which accounts for the large base exchange capacity of this mineral.

Other montmorillonite types of minerals. There are a number of other minerals similar to montmorillonite that are found in some clays and soils. While these minerals are of great interest to the soil chemist, they do not play an important role in ceramics.

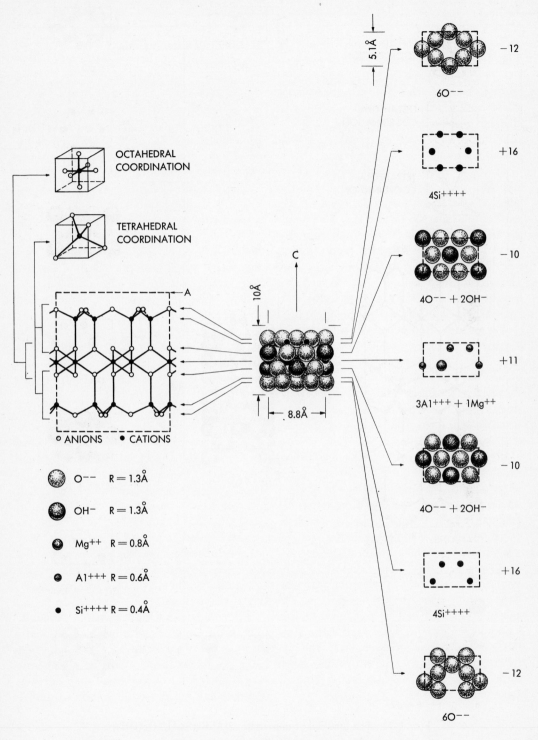

F$_{\text{IG}}$. 1–6. The structure of montmorillonite, $2[(\text{Al}_{1.67}\text{Mg}_{0.33})\ \text{Si}_4\text{O}_{10}(\text{OH})_2]$.

11

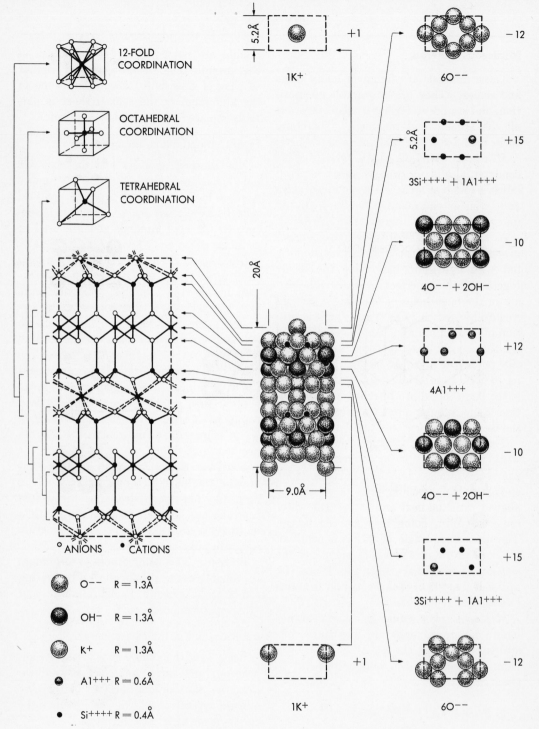

12-FOLD COORDINATION

OCTAHEDRAL COORDINATION

TETRAHEDRAL COORDINATION

° ANIONS • CATIONS

O^{--} R = 1.3Å

OH$^-$ R = 1.3Å

K$^+$ R = 1.3Å

Al^{+++} R = 0.6Å

Si^{++++} R = 0.4Å

5.2Å

1K$^+$ +1

20Å

9.0Å

6O^{--} −12

3Si^{++++} + 1Al^{+++} +15

5.2Å

4O^{--} + 2OH$^-$ −10

4Al^{+++} +12

4O^{--} + 2OH$^-$ −10

3Si^{++++} + 1Al^{+++} +15

1K$^+$ +1

6O^{--} −12

FIG. 1–7. The unit cell of muscovite, $2[Al_4K_2(Si_6Al_2) O_{20}(OH)_4]$.

12

Micaceous minerals

These minerals are found in many clays and shales. They are of variable composition, and studies so far have not worked out all the specific end members. Therefore, there is an excellent field for research here.

Muscovite [$Al_4K_2(Si_6Al_2)O_{20}(OH)_4$]. This common mica is not in itself a clay mineral but it is found as an accessory in many clays and forms an end member for the hydrated micas with claylike properties.

As shown in Fig. 1–7, muscovite consists of sheets similar to those of montmorillonite, but with potassium ions tying the sheets together into a perfect crystal. The K^+ are in the open holes of the Si-O layer and weakly bonded to 12 O^{--}'s. The substitution of an aluminum ion for one of silicon in the tetrahedral layer balances the charge of the potassium ion. It should be remembered that even though the aluminum ion is too large to go into tetrahedral coordination alone, it may substitute for silicon in the tetrahedral layer to a limited extent.

The bonding of the potassium ions is weak, which accounts for the perfect cleavage of this mineral, but the bonds are strong enough to cause each layer to be accurately aligned with the next and to prevent water molecules from going between the layers. The monoclinic angle of 95° is a result of the shift in the octahedral layer, not of the shifting of one sheet over another. The unit cell consists of two sheets and contains two molecular weights.

Micaceous clay minerals. These rather indeterminate minerals are important constituents of many clays. They have been named *illites* by Grim and correspond to the formula:

$$Al_{4-a+b}Mg_aFe_cK_2Si_{8-y}Al_yO_{20}OH_4.$$

Sericite, hydromica, bravaisite, and brommallite are names applied to members of this group. The structure is very similar to that of muscovite except that some of the aluminum in the central sheet is partially replaced by magnesium and iron.

Hydrated aluminous minerals

The minerals gibbsite and diaspore are not considered clay minerals by many workers in the field, but they are included here because they are analogous in structure to the clay minerals and are used in ceramics as clays.

Gibbsite. This mineral has the formula $Al(OH_3)$ and consists of a simple sheet structure as shown in Fig. 1–8. The aluminum is in octahedral coordination with the hydroxyl groups, but only $\frac{2}{3}$ of the available positions are filled. This is the same structure found in the octahedral sheet of the other clay minerals. The bonds are weak; thus the mineral is soft and breaks up on heating with relative ease. Gibbsite is a common constituent of soils and clays in warm climates, and is found in a few deposits of high purity.

Diaspore ($HAlO_2$). This is an important mineral in the refractories industry. It is probable that the hydrogen is in twofold coordination between two oxygens with a hydroxyl bond making up a sheet structure somewhat like that of gibbsite.

Bauxite. This common ore of aluminum is probably not a specific mineral but rather a mixture of gibbsite, kaolin, limonite, and other minor minerals.

Identification of the clay minerals

Because of their small size, variable composition, and often imperfect crystallization, the positive identification of the clay mineral crystals is difficult and requires the employment of all possible means.

Petrographic methods. In cases where

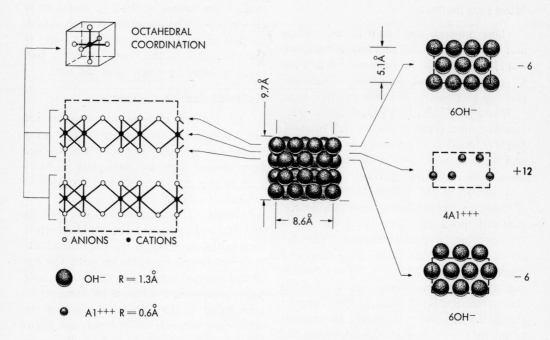

FIG. 1–8. The unit cell of gibbsite, 8[Al(OH)₃].

Table 1-5

Optical, Crystallographic, and Thermal Properties of Some Clay Minerals

Properties	Kaolinite	Montmorillonite	Gibbsite	Muscovite
Np	1.560	1.532	1.567	1.552
Ng	1.566	1.557	1.589	1.558
Ng-Np	0.006	0.025	0.022	0.036
Crystal Class	Monoclinic	---	Monoclinic	Monoclinic
a	5.14Å	5.10Å	5.064Å	5.2Å
b	8.90Å	8.83Å	8.620Å	9.0 Å
c	14.51Å	10.Å	9.699Å	20Å
β	100°12'	---	85°29'	89°54'
Endothermic effect	450°C	100°, 600°, 850°C±	250°C	small
Exothermic effect	980°C	700°-800°	---	small
Specific gravity	2.0-2.5	2.0-2.5	2.3-2.4	2.76-3.00

Np = least index of refraction
Ng = greatest index of refraction
Ng-Np = birefringence
a,b,c = dimensions of the unit cell along the x-, y-, and z-axes respectively
β = angle of the c-axis with the a-axis

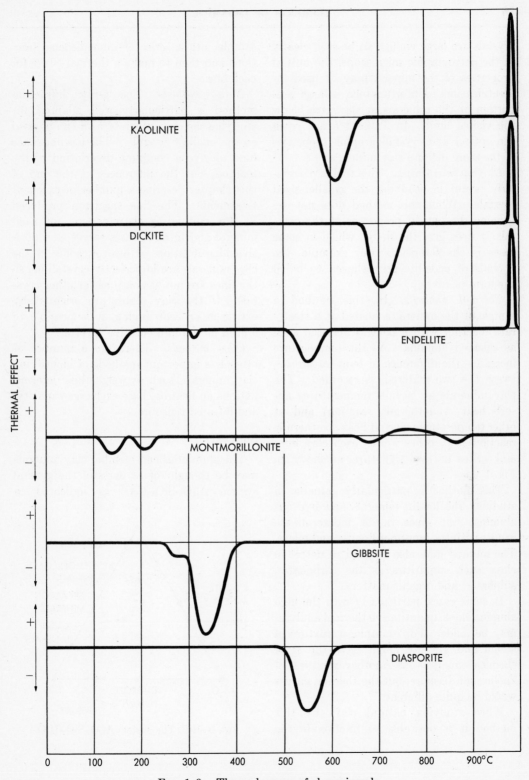

FIG. 1–9. Thermal curves of clay minerals.

15

crystals are large enough to be seen clearly in the petrographic microscope, the optical properties of the mineral may be used for identification. Unfortunately, a large proportion of the minerals in clays are below the visible limit. In Table 1–5 are given the optical and crystallographic properties of the more definite clay minerals.

Electron microscope. This tool is particularly useful in studying the smaller-sized crystals. While this method does not determine the optical properties of the crystals, it does give the shape, which in some cases is characteristic. For example, the crystals of endellite are believed to be in the form of tubes.

Thermal analysis. In this method a sample of the mineral is heated at a steady rate together with a neutral material such as calcined alumina. As the temperature increases, the difference in temperature between the two materials is recorded. The clay minerals at certain temperatures absorb heat (endothermic reaction) and at other temperatures give off heat (exothermic reaction). In other words, each clay mineral writes its own signature, as shown in Fig. 1–9.

This method is particularly valuable in studying the kaolin minerals and hydrous alumina, but gives rather indeterminate results with the montmorillonites and micas. The method may also be used to detect in clays such impurities as the carbonates, sulphates, and organic matter.

In some cases, particularly with the high alumina clays, quantitative thermal analyses may be made. For example, a mixture of kaolin and gibbsite might have the same chemical analysis as another mixture of kaolin and diaspore, but the thermal curves would be quite different.

When studying natural clays by this method it is desirable to fractionate the sample into a series of monodisperse fractions and then to make a thermal curve for each one.

X-ray methods. The x-ray diffraction method is particularly well adapted for studying the clay minerals, as it can be used on the smallest crystals. There is not space here to give a complete description of the method, but the references at the end of this chapter describe a number of excellent treatments. The line spectrum produced in this method is characteristic for each mineral and in most cases serves as a positive identification. However, some of the clay minerals are imperfectly crystalline, so the lines are not as sharp as for other crystals. If the clay plates are oriented by extrusion or centrifuging, it is possible to obtain more representative spectra.

Other methods. There are a number of other less important methods of identifying clay minerals, such as weight loss determinations on heating, base exchange capacity, and infrared reflection.

Summary.

The aluminous end-member clay minerals may be thought of as parts of the general system Al_2O_3-SiO_2-H_2O, as indicated in

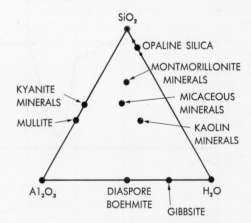

FIG. 1–10.   The system Al_2O_3-SiO_2-H_2O.

Fig. 1–10. Such a system is difficult to work out because of the high pressures involved and the sluggishness of the reactions; however, progress has been made and eventually a complete picture of the system will be available.

REFERENCES

Crystal Chemistry

BRAGG, W. L., *Atomic Structure of Minerals*, Cornell University Press, Ithaca, N. Y., 1937.

EVANS, R. C., *An Introduction to Crystal Chemistry*. Cambridge University Press, London, 1938.

RIGBY, G. R., "The Applications of Crystal Chemistry to Ceramic Materials." *Trans. Brit. Ceram. Soc.*, **48**, 1, 1949.

STILLWELL, C. W., *Crystal Chemistry*. McGraw-Hill Book Co., Inc., New York, 1938.

Crystallography

PHILLIPS, F. C., *An Introduction to Crystallography*. Longmans, Green and Co., Inc., New York, 1946.

Structure of Clay Minerals

DAVIS, D. W., ROCHON, T. G., ROWE, F. G., et al., "Electron Micrographs of Reference Clay Minerals." *Am. Petroleum Inst. Proj. 49*, Prel. Rep. No. 6, 1950.

EWING, F. J., "The Structure of Diaspore." *J. Chem. Phys.* **3**, 203, 1935.

GRUNER, J. W., "The Crystal Structure of Kaolinite." *Z. Krist.* **83**, 75–88, 1932.

HOFMAN, N., ENDELL, K., and WILM, D., "Kristall Structur und Quellung von Montmorillonit." *Z. Krist.* **86**, 340, 1933.

JACKSON, W. W., and WEST, J., "The Crystal Structure of Muscovite." *Z. Krist.* **76**, 211, 1930–1931.

KERR, P. F., and HAMILTON, P. K., "Glossary of Clay Mineral Names." *Am. Petroleum Inst. Proj. 49*, Prel. Rep. No. 1, 1949.

KERR, P. F., MAIN, M. S., and HAMILTON, P. N., "Occurrence and Microscopic Examination of Reference Clay Mineral Specimens." *Am. Petroleum Inst. Proj. 49*, Prel. Rep. No. 5, 1950.

MEGAW, H. D., "The Crystal Structure of Hydrargillite, $Al(OH)_3$." *Z. Krist.* **87**, 185, 1934.

ROSS, C. S., and HENDRICKS, S. B., *Minerals of the Montmorillonite Group*. U. S. Geol. Survey, Prof. Paper 205–B, 23, 1945.

ROSS, C. S., and KERR, P. F., *Halloysite and Allophane*. U. S. Geol. Survey, Prof. Paper 185–G, 135–148, 1934.

ROSS, C. S., and KERR, P. F., *The Kaolin Minerals*. U. S. Geol. Survey, Prof. Paper 165, 151–180, 1931.

Identification Methods

BUERGER, M. J., *X-ray Crystallography*. John Wiley and Sons, Inc., New York, 1942.

NORTON, F. H., "Critical Study of the Differential Thermal Method for the Identification of Clay Minerals." *J. Am. Ceram. Soc.* **22**, 54, 1939.

RIGBY, G. R., *The Thin-Section Mineralogy of Ceramic Materials*. Brit. Ref. Res. Assn., Stoke-on-Trent, England, 1948.

ROGERS, A. F., and KERR, P. F., *Optical Mineralogy*, 2nd ed. McGraw-Hill Book Co., Inc., New York, 1942.

WINCHELL, A. N., *Elements of Optical Mineralogy*, Parts I, II, and III, 5th ed. John Wiley and Sons, Inc., New York, 1937.

CHAPTER 2

ORIGIN AND OCCURRENCE OF CLAYS

Introduction

Clays are so widespread that a deposit of some type or other is found in nearly every county in the United States. The clays vary in character over a wide range; some are particularly valuable to the ceramic industry, while others are so impure that they cannot be used in fired products. Some deposits of clay are found in the same positions as the parent rocks from which they were derived, while others have been deposited at great distances from their point of origin. It is the purpose of this chapter to describe briefly the different types of clay, their locations, and how they were formed.

Classification of clays

It may serve best to orient the reader if a classification of clays is set up. Any classification depends on the point of view of the classifier; if he is a geologist, then a classification as to origin would seem most logical to him; on the other hand, a producer of clay ware would be more interested in a classification based on the properties of the clay. Therefore, two classifications will be used here, one based on the method of formation in the earth and the other based on the use to which the clay is put in the ceramic industry.

As geologic terms must be used occasionally from here on to describe formations in the earth's crust, a time scale is shown in Fig. 2–1 to refresh the memory of those whose knowledge of geology is a bit rusty.

In Table 2–1 is given a classification of clays arranged by Stout. It will be seen that there are two general classes, residual clays and transported clays.

In Table 2–2 there is shown a classification based on use, but such a classification cannot be too rigid.

Origin of clays

In general, clays are a secondary product in the earth's crust produced by the alteration of rocks of the pegmatite type.

Residual clays. A clay that is found in the same position as the parent rock from which it was derived is called residual. It is now generally believed that most residual clays are formed by a series of reactions caused by percolation of ground water through the mass, aided by many other weathering factors such as freezing. This water contains dissolved CO_2 from the air and organic acids from vegetation. In the following equations the steps in the process are delineated. However, there is no evidence that in nature such a step by step process occurs. Rather, many of the steps take place simultaneously. Also, there is good evidence for considerable migration by chemical and colloidal transfer from one part of the deposit to the other.

(1) $\underline{KAlSi_3O_8 + H_2O}$
   feldspar
   $\longrightarrow HAlSi_3O_8 + KOH$ (hydrolysis)

(2) $HAlSi_3O_8 \longrightarrow$
   $(OH)AlSi_2O_5 + SiO_2$ (desilication)
   pyrophyllite

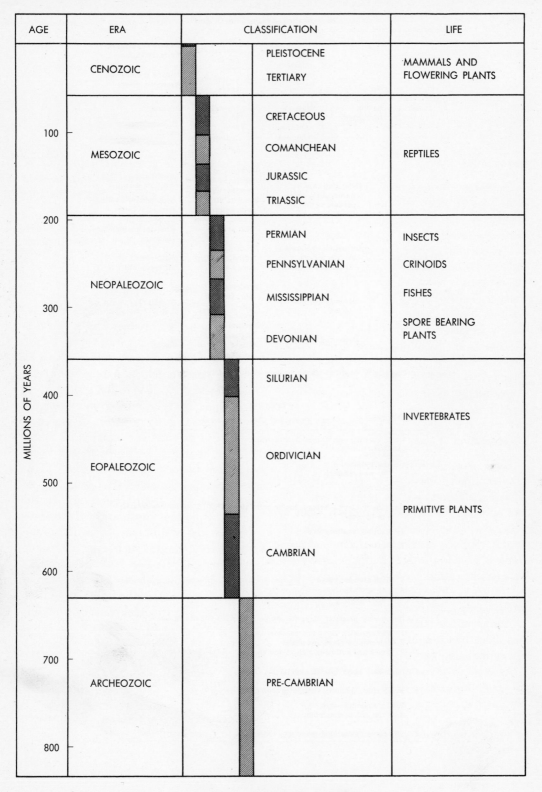

AGE	ERA	CLASSIFICATION		LIFE
	CENOZOIC		PLEISTOCENE	MAMMALS AND FLOWERING PLANTS
			TERTIARY	
100	MESOZOIC		CRETACEOUS	REPTILES
			COMANCHEAN	
			JURASSIC	
200			TRIASSIC	
	NEOPALEOZOIC		PERMIAN	INSECTS
			PENNSYLVANIAN	CRINOIDS
300			MISSISSIPPIAN	FISHES
			DEVONIAN	SPORE BEARING PLANTS
400	EOPALEOZOIC		SILURIAN	INVERTEBRATES
500			ORDIVICIAN	
				PRIMITIVE PLANTS
600			CAMBRIAN	
700	ARCHEOZOIC		PRE-CAMBRIAN	
800				

FIG. 2–1. Geologic time scale.

19

Table 2-1

Classification of Clays as to Origin*

Residual matter	No movement during formation	Products of ordinary weathering	From crystalline rocks → Impure residual clay / Primary kaolin
			From sedimentary rocks → Impure residual clay / Kaolinitic clay
		Same as above with additional chemical action	From crystalline rocks → Bauxite
			From sedimentary rocks → Bauxite / Diaspore
Transported matter	Deposited in still water, little or no current action, seas, lakes, bogs, etc.	Products of ordinary weathering	Argillaceous shale / Argillaceous silt
		Same as above with additional intensive chemical action	Sedimentary kaolin / Ball clay / Some bauxite / Coal-formation clay / Diaspore
	Deposited by slowly moving waters, streams, estuaries, etc.	Products of grinding with some weathering	Siliceous shale / Siliceous silt
	Deposited by glacial action	Products of abrasion with slight weathering	Glacial clay or till
	Deposited by winds	Products of abrasion with slight weathering	Loess

*Lecture on Clays by W. Stout, at Massachusetts Institute of Technology, 1937.

Table 2-2

Classification of Clays as to Use

A. White burning clays (used in whiteware)

    1. Kaolins
       a. residual
       b. sedimentary

    2. Ball clays

B. Refractory clays (having a fusion point above $1600^{\circ}C$ but not necessarily white burning)

    1. Kaolins (sedimentary)

    2. Fire clays
       a. flint
       b. plastic

    3. High alumina clays
       a. gibbsite
       b. diaspore

C. Heavy clay-products clays (of low plasticity but containing fluxes)

    1. Paving brick clays and shales
    2. Sewer-pipe clays and shales
    3. Brick and hollow tile clays and shales

D. Stoneware clays (plastic, containing fluxes)

E. Brick clays (plastic, containing iron oxide)

    1. Terra-cotta clays
    2. Face and common brick

F. Slip clays (containing more iron oxide)

(3) $HAlSi_3O_8$
$$\longrightarrow HAlSiO_4 + 2SiO_2 \quad \text{(desilication)}$$

(4) $2HAlSiO_4 + H_2O$
$$\longrightarrow \underline{(OH)_4Al_2Si_2O_5} \quad \text{(hydration)}$$
$$\text{kaolinite}$$

(5) $HAlSiO_4$
$$\longrightarrow \underline{HAlO_2} + SiO_2 \quad \text{(desilication)}$$
$$\text{diaspore}$$

(6) $HAlO_2 + H_2O$
$$\longrightarrow \underline{Al(OH)_3} \quad \text{(hydration)}$$
$$\text{gibbsite}$$

The opalescence of the water in brooks emerging from residual kaolin deposits is proof that desilication actually takes place.

There are some residual clay deposits, notably those of Cornwall, which are so deep below the surface that alteration by ground water seems unlikely. In these cases the agent is believed to be fluorine-containing gases rising through the rock.

The purity of the residual clay depends on the purity of the parent rock, the completeness of alteration, the amount of impurities removed, and the amount of impurities brought into the deposits. Even the purest residual kaolins contain considerable unaltered rock, and it is not unusual to obtain only ten per cent pure clay from a deposit.

Some residual clays are believed to have been derived from the solution of limestones originally containing clay impurities. Others come from disintegration of shales.

Gibbsite or diaspore may be formed as a residual product of weathering from a feldspathic parent. In this case the weathering process goes to an end point which is aided by porosity of the deposit and a warm climate.

Sedimentary clays. The sedimentary kaolins of the southern United States are noteworthy because of their uniformity and

great extent, running close to the fall line from New Jersey to Mississippi.

These kaolins are formed mainly from silts washed down into lakes and lagoons from the higher levels. While some kaolization may have taken place before and during transportation, the main alteration occurred after the sediment had been laid down by the same general process described for residual clays. These sedimentary deposits contain much less sand and fewer rock fragments than the residual clays because, by natural classification, only the finer fractions were laid down.

Ball clays are sedimentary in origin, are usually found in swamps, and have a high amount of organic matter. The alteration of the fine sediments undoubtedly was accelerated by organic acids.

Coal formation clays are derived from silts laid down in ancient swamps in which the coal vegetation grew. There has been much discussion as to the reason for the high purity of some of these clays. Theories advanced are that the clay is (1) the altered material from the oxidation product of the coal vegetation, (2) the altered soil in which the coal vegetation grew and which thus was depleted of alkalis, (3) the altered soil leached by strong plant acids.

Glacial clays are widespread on the eastern seaboard and serve as material for brick and other heavy clay products. They are derived from the rock dust washed out of the glaciers, which is then deposited in lakes or estuaries where the alteration takes place. They are rather impure and often contain stones, gravel, or sand.

Shales are of sedimentary origin and are variable in composition. They are hard and often contain the mineral sericite. Although some are quite pure, most of them contain considerable amounts of iron minerals.

Laboratory synthesis of clay minerals.
Most of the clay minerals have been made
in the laboratory under hydrothermal conditions. The final mineral depends on the
starting material, the pressure, the temperature, and the acidity of the environment.

Kaolins

Residual kaolins. The map in Fig. 2–2
shows the location of some of the deposits
of residual kaolin in this country. It will
be seen that they run in a band from Ver-
mont to Georgia and up the Mississippi
Valley, with some scattered deposits in the
West. The most important deposits are
located near Spruce Pine in North Carolina.
The partially altered rock is mined, then
put through a careful washing process to
produce a pure, uniform product.

Residual kaolins derived from Oriskany
limestones have been worked in Pennsylvania, and a deposit is worked in the state
of Washington.

The residual kaolins of Cornwall in England are world famous as the source of
"English China Clay." The clay is washed
out of the deposit with a high pressure
stream of water and allowed to collect with
the sand in a sump. The suspended clay
is pumped up to the ground level and
treated as described in the next chapter.

Other famous deposits of residual kaolin
are at Zettlitz in Germany, Karlsbad in
Czechoslovakia, and in the places shown in
the map of Fig. 2–3.

Sedimentary kaolins. In Fig. 2–4 are
shown the deposits of sedimentary kaolins
in this country. The important workings
are of Cretaceous age and are located in
South Carolina, Georgia, and Florida.

Most of the deposits are covered with a
gravel over-burden which is stripped off by
power shovels or modern earth-moving

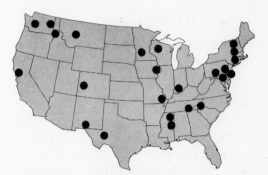

Fig. 2–2. Residual kaolin deposits in the
United States.

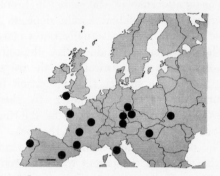

Fig. 2–3. Residual kaolin deposits in Europe.

Fig. 2–4. Sedimentary kaolin deposits in the
United States.

machinery. The kaolin, usually rather
soft, may be taken out in the same way,
although some companies are using a shale
planer. In one case the clay is blunged
(beaten up in water) into a slip at the point

of mining and pumped to the processing plant.

The Florida kaolins occurring in Tertiary (Eocene) deposits have a finer grain size than others and therefore are a valuable ingredient in many whiteware bodies. These clays are dredged out of lakes and washed in much the same way as other kaolins.

A large proportion of the sedimentary kaolin goes into paper fillings and coatings as well as other fillers. The ceramic industry uses it for whiteware bodies and in refractories.

Ball clays

As shown in the map of Fig. 2–5, the deposits of this clay are found in a rather restricted area in this country. The most

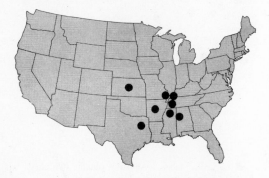

FIG. 2–5. Ball clay deposits in the United States.

FIG. 2–6. Ball clay deposits in Europe.

important deposits are in western Kentucky and western Tennessee. These Tertiary formations occur in large lenses with a variety of clay in different strata. There are excellent deposits of ball clay in England at Devonshire (Fig. 2–6), but few good deposits occur in continental Europe, unless the glass pot clays of Germany may be considered in this classification.

The ball clays in this country are mined with earth-moving equipment, partially dried, and shipped. No serious attempt has been made by the producers to supply water-washed ball clay because of the difficulty of filtering. However, air-floated clays are produced by disintegration and removal of the coarse particles.

Ball clays are used almost exclusively in whiteware bodies to increase strength and plasticity. They are little used in such bodies as hard porcelain, frit porcelain, or bone china, as they reduce the translucency.

Fire clays

Flint fire clays. In this country we are fortunate in having ample supplies of flint fire clay, a product not so readily available to the European manufacturer. These clays are hard; they break with a conchoidal fracture and develop little plasticity, even after grinding. However, mixed with plas-

FIG. 2–7. Flint fire clay deposits in the United States.

tic clay they act as a grog (highly burned clay) and maintain the volume stability of the firebrick made from them.

The map in Fig. 2–7 shows the rather limited areas where this clay is found, the most important deposits being in Pennsylvania, Ohio, Maryland, and Kentucky. Many of the deposits occur in the carboniferous strata with coal, so much of the mining is underground. In some cases coal and clay are mined together.

Plastic fire clays. Plastic fire clay, often of Tertiary age, is widespread in this country, as shown in the map of Fig. 2–8. The composition varies considerably, as will be shown in the next chapter. The aluminous clays of Pennsylvania are at one end of the scale and the siliceous clays of New Jersey at the other. Each clay, however, has a use which it serves best in the refractories industry.

There is also much fire clay for low heat duty that contains considerable flux. Some of these clays are excellent for ladle brick and plastics, and may also be used for other ceramic products. Their widespread occurrence is shown in Fig. 2–9. Some firebrick are made entirely of plastic clay, others are mixtures of plastic clay and flint clay or grog. Much of the plastic fire clay is used in mortars, plastics, and castables.

High alumina fire clays. These clays are important in making super-duty and high alumina firebrick for severe service. The chief source is the diaspore in the Ozark region of Missouri, as shown on the map of Fig. 2–10. These clays occur in pockets and are thought to be altered from flint clay by percolating water. Much careful selection is required to maintain a uniform material.

Another source for high alumina refractories is the bauxites of the South, shown on the map of Fig. 2–11. These are not as suitable as diaspore because they shrink excessively in firing, even though they may have the same chemical analysis.

Calcareous and ferrogeneous clay

These clays contain too much fluxing ma-

FIG. 2–9.   Low heat duty plastic fire clay deposits in the United States.

FIG. 2–8.   High heat duty plastic fire clay deposits in the United States.

FIG. 2–10.   Diaspore deposits in the United States.

terial for refractories and too much iron for whiteware use. Often they are from carboniferous shale in the central states. They are, however, very useful in stoneware and heavy clay products where strong structures are produced at moderate firing temperatures. These clays are widespread, as shown in Fig. 2–12. More detailed information on the locations is available in the "Report of the Committee on Geological Surveys of the American Ceramic Society," *Bull. Am. Ceram. Soc.* **24,** 234, 1945.

Glacial clays. These red-burning brick clays are found extensively on the eastern seaboard and serve for face and common brick, as well as flowerpots and garden pottery.

Fig. 2–11. Bauxite deposits in the United States.

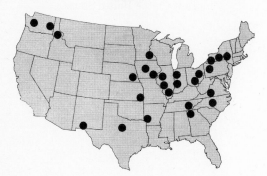

Fig. 2–12. Paving brick and sewer pipe clay deposits in the United States.

Special clays

Allophane. This soapy, amorphous clay is of limited occurrence, and small pockets are often associated with halloysite. No commercial deposit is known.

Halloysite. A soapy white mineral, halloysite, has been found in many places. Some possible commercial deposits are found in Indiana and North Carolina. This mineral, because of its whiteness and highly plastic properties, may become of value to the whiteware industry.

Slip clays. These highly impure clays are usually of glacial origin. They are found in a number of places, but those from Albany, New York are best known. These clays, melting down at about 1200°C, are used on stoneware and electrical insulators as a glaze.

Stoneware clays. These clays contain some feldspar and vitrify into a dense body. They are found in many places (Fig. 2–12), particularly in the Carboniferous strata of the central states and the Tertiary strata of the East and West. High grade stoneware is now made of a compounded body rather than from a single clay.

Bentonite. This clay is the decomposed product of volcanic glass and is found in many places in the western United States,

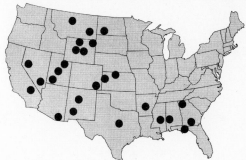

Fig. 2–13. Bentonite deposits in the United States.

as shown on the map in Fig. 2–13. Some bentonites swell greatly when wet, while others show little swelling. The main use of this material in ceramics is as a plasticizer in refractory mixes.

Production and prices of clay

Because much of the lower-grade clay and some of the refractory clay is mined and used in the same organization, it is difficult to get complete figures on production, since the published statistics apply only to clay sold from the mine. However, Table 2–3 gives an estimate of the clay produced in the United States, together with its price at the mine. U. S. Bureau of Mines figures show that only 112,400 short tons of clay were imported in 1947, the majority of it

Table 2-3

Production and Prices of Clay in The United States

Type of clay	Production in 1946 (Millions of tons)	Price in dollars per ton
Ball clay	0.2	{ crude 6-8 air-floated 15-17
Kaolin[1]	1.3	10
Fire clay[2]	7.9	2.5
Bauxite[3]	2.3	7
Bentonite	0.6	12
Heavy clay-products clays[4]	25.0	0.5-3.0[4]

[1]Air-floated or washed.
[2]This does not include all produced.
[3]Approximately 15% of this used in ceramics.
[4]Estimated.

being English china clay for the paper trade. This is a very small fraction of our own production.

REFERENCES

CHELIKOWSKY, J. R., "Geologic Distribution of Fire Clays in the United States." *J. Am. Ceram. Soc.* **18**, 367, 1935.

Minerals Year Book. U. S. Bur. of Mines, Washington, D. C., 1951.

MOREY, G. W., and INGERSON, E., "The Pneumatolytic and Hydrothermal Alteration and Synthesis of Silicates." *Econ. Geol.*, Supp. to Vol. XXXII, **5**, 607, 1937.

REIS, H., *Clays, Occurrence, Properties and Uses*, 3rd ed. John Wiley and Sons, Inc., New York, 1927.

REIS, H., "Report of Committee on Geological Surveys." *Bull. Am. Ceram. Soc.* **24**, 234, 1945.

REIS, H., and BAILEY, W. S., "High Grade Clays of the United States." *U. S. Geol. Survey Bull.*, p. 708, 1922.

Ross, D. W., "Nature and Origin of Refractory Clays." *J. Am. Ceram. Soc.* **10**, 704, 1927.

CHAPTER 3

PROPERTIES OF CLAYS

Introduction

As clay is the chief raw material used in ceramics, no apology is needed for devoting a whole chapter to a discussion of its properties. This is particularly true because there is no one publication in which the properties of ceramic clays have been adequately treated.

Properties of the raw clay

Particle size. The particle size of clay is a very important characteristic, since it influences many other properties such as plasticity, dry strength, and base exchange capacity.

There are numerous ways of measuring particle size. The microscope and the electron microscope yield absolute size values, but to measure a large number of particles with them is indeed a tedious process. The x-ray method is excellent for particles less than 0.1 micron in size, but requires careful work to measure the resultant broadening of the lines on the diffraction pattern. The most feasible method for the average analysis is the sedimentation method, where the rate of settling of the particles in water is measured and then converted by Stokes' law into particle size. The assumption must be made that the particles are individual and that their settling rate is the same as that of equivalent spheres. This requires a dilute suspension and complete deflocculation. It has been shown that the platelike clay particles do settle at almost the same rate as a sphere having the same diameter as the width of the plate. Static settling becomes so slow when the particles reach one micron or less in size that thermal currents and Brownian movement tend to introduce serious errors. Therefore the measurements below this size are made in a centrifugal field of force. More details about methods can be found in the references at the end of the chapter.

There are numerous methods for plotting the results of the particle size analysis, but it has been found that for clays a simple plot of per cent finer versus the logarithm of size is the most convenient. In Fig. 3–1 there are shown particle size curves for a number of ceramic clays. It will be seen that there is a great difference between them. Over half the weight of the particles of Kentucky ball clay have a size below one micron, while in Pennsylvania fire clay only one-fifth are below this limit. Many clays have little or no portion below 0.1 micron, while some have several per cent in this range. It should be remembered that the most active portion of the clay is in the fine range because of the enormous surface area. Particle size measurements that do not carry down to the finest end of the scale have little meaning in evaluating a clay.

Particle shape. Very little is known about the exact shape of clay particles, but the electron microscope will be able to give us this information. For example, the kaolinite plates in Fig. 1–4 show the outline clearly, and the shadow at the edge indicates the thickness, which is one-third of

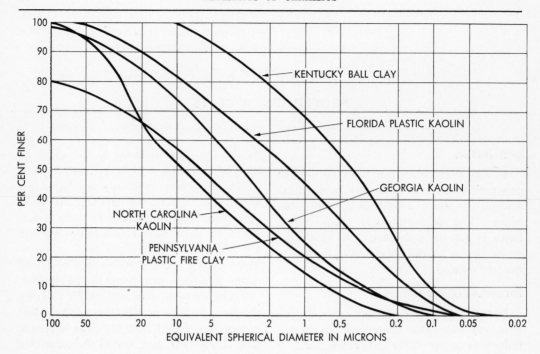

FIG. 3–1. Particle size distribution of several natural clays.

the shadow width. In general, kaolins seem to have particles with a thickness of 8 to 10 per cent of the width. How closely this holds for wide ranges of particle size or for different kaolin sources is something that must be determined. A slight change in particle thickness will make a great change in the total surface area of a gram of clay.

Some of the clay minerals, for example endellite, seem to be made up of tubes rolled up from a single, very thin crystal.

Other clay minerals are not as smooth as kaolinite, but are somewhat ragged. Gibbsite particles have actually a rather spongy appearance.

Base exchange capacity. It was shown in Chapter 1 that when a clay mineral such as montmorillonite, with a balanced lattice, had some of the ions replaced by others of different valence (for example, Al^{+++} replaced by Mg^{++}), there would be set up a

deficiency of charge in the structure as a whole. This deficiency is balanced by ions adsorbed on the surface of the crystal. However, this is not the only way that ions may be adsorbed, for a balanced lattice like kaolinite can adsorb a small number of ions. It has been thought that this adsorption is due to the broken bonds at the edges of the crystal. This is illustrated diagrammatically in Fig. 3–2, where there are 20 broken bonds for one unit cell of kaolinite.

If a monodisperse (particles of one size) fraction of kaolinite is available, it is possible to calculate the number of broken bonds per crystal in relation to the ions adsorbed. It will be instructive to make such a calculation for a specific kaolinite particle. The continual rejection of superfluous figures in calculations of this kind should be noted.

The kaolinite crystal is a hexagonal plate 6.0×10^3 Å across the points, and 5.0×10^2

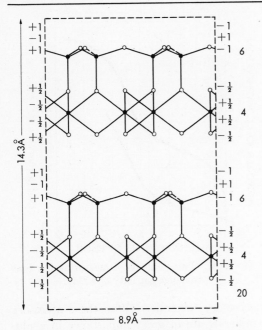

FIG. 3–2. Schematic diagram of unit cell of kaolinite, showing the broken bonds.

Å thick. It is easy to see that the edge area will be

$$6 \times 3.0 \times 5.0 \times 10^5 = 9.0 \times 10^6 \text{ sq Å.}$$

The volume of each particle will be

$$24 \times 10^6 \times 5.0 \times 10^2$$
$$= 12 \times 10^9 \text{ cu Å} = 12 \times 10^{-15} \text{ cc.}$$

The weight of each particle is then

$$12 \times 10^{-15} \times 2.6 = 31 \times 10^{-15} \text{ gm.}$$

Turning to the unit cell shown in Fig. 3–2, the average area of a side is

$$\frac{8.9 + 5.1}{2} \times 14.3 = 100 \text{ sq A.}$$

The number of broken bonds per unit cell may be seen in the same figure to be 12 from the Si^{++++} and 16 half-strength bonds from the Al^{+++}, making a total of 20, or for the average face, 5.0.

Then the number of broken bonds on the edge of one kaolinite particle will be

$$\frac{9.0 \times 10^6}{100} \times 5.0 = 4.5 \times 10^5.$$

The base exchange capacity of one gram of this kaolinite is found by experiment to be 1.6×10^{-5} equivalents per gram. Multiplying by Avogadro's constant, 6.0×10^{23}, and by the weight of one particle gives the number of ions adsorbed on one particle:

$$1.6 \times 10^{-5} \times 6.0 \times 10^{23} \times 31 \times 10^{-15}$$
$$= 3.0 \times 10^5.$$

This close agreement is interesting but not conclusive, for some of the ions may be adsorbed on the face of the crystal, and some of the edge positions may be vacant.

The maximum capacity to adsorb ions is called the base exchange capacity and is expressed in milliequivalents per 100 grams of clay.

There are a number of methods of meas-

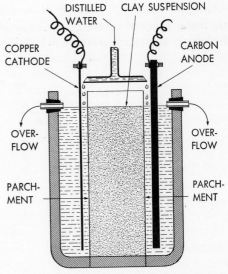

FIG. 3–3. Three-compartment electrodialysis cell.

uring this quantity, but in ceramics it is convenient to electrodialyze the clay slip in a three-compartment cell, as shown in Fig. 3–3. In this way the adsorbed ions are stripped off and replaced by hydrogen. If the hydrogen-clay slip is then titrated with a base such as NaOH, and the *pH* (hydrogen ion concentration) of the slip measured, a curve like that in Fig. 3–4 will be obtained. The inflection point in this curve will represent the base exchange capacity.

Another method is to measure the electrical conductivity of the hydrogen-clay slip as a base is added, giving a curve also shown in Figure 3–4, where the intersecting straight lines represent the base exchange capacity.

In Table 3–1 are listed the base exchange capacities for some monodisperse fractions of kaolinite and for a few complete clays. For kaolinite, the base exchange capacity is closely proportional to the total surface area.

As will be shown in Chapter 10, the base exchange capacity is an important property of casting slips.

Accessory minerals. Natural clays contain many types of crystalline matter, and all that are not clay minerals are known as

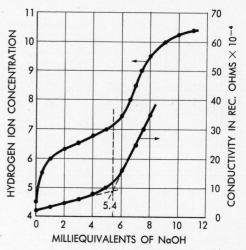

Fig. 3–4. Two methods of measuring the base exchange capacity of a clay.

accessory minerals. Often these accessory minerals are of great importance in evaluating the worth of a clay. These minerals can be determined most readily by centrifuging the clay in a series of heavy liquids to separate the various minerals in groups according to density. A suitable division is shown in Table 3–2. After each group is separated, it may be examined under the petrographic microscope or by x-ray diffrac-

Table 3-1

Base Exchange Capacity

Clay	Mean spherical diameter in microns	Surface area per 100 gm/m²	Base exchange capacity in milli-equiv per 100 gm
Kaolinite fraction	10	1.1	0.4
" "	4	2.5	0.6
" "	2	4.5	1.0
" "	1	11.7	2.3
" "	0.5	21.4	4.4
" "	0.2	39.8	8.1
Georgia kaolin			1
Pa. flint fire clay			5
Ky. flint fire clay			7
Plastic fire clay			7
Ky. ball clay			12
Bentonite			100

Table 3-2

Separation of Accessory Minerals in Clay

Mineral	Specific gravity	Heavy liquid
Gypsum	2.3	
Gibbsite	2.4	2.5-Tetrabromoethane + carbon tetrachloride
Orthoclase	2.6	
Microcline	2.6	
Albite	2.6	(Clay minerals in this group) 2.65-Tetrabromoethane + carbon tetrachloride
Quartz	2.7	
Calcite	2.7	
Anorthite	2.8	
Biotite	2.8	
Muscovite	2.8	
Beryl	2.8	2.9-Tetrabromoethane + carbon tetrachloride
Tourmaline	3.0	
Magnesite	3.0	
Garnet	3.4	
Limonite	3.8	3.87-Sol. of thallium formiate at 40°C
Corundum	4.0	
Ilmanite	4.2	
Rutile	4.5	
Zircon	4.7	
Pyrite	5.0	
Hematite	5.2	
Magnetite	5.2	

tion methods. There the species and relative amounts of the minerals can readily be determined, since generally there are only 2 or 3 minerals in each group.

The more common accessory minerals in clays are quartz, feldspars, micas, and the iron minerals.

Organic matter. All clays contain some organic matter. There is very little in the residual kaolins but a large amount in ball clays. The organic matter is in the form of lignite, waxes, or humic acid derivatives. No comprehensive work has been carried out in this field, with the exception of the paper by Sharratt and Francis, although the organic matter undoubtedly has an important influence on the plastic and dried properties of clays.

In studying the properties of the clay minerals it is necessary to remove the organic matter without destroying the crystal structure. This may most readily be done by digestion in a hydrogen peroxide solution. A long time is required to remove all organic matter; several months, with frequent renewal of the peroxide, is often necessary.

Color. The color of the raw clay is of little importance in ceramics, where heat will destroy or alter it. For the paper trade, however, color is of great importance, and it is possible to obtain an exact measure of it by means of the recording spectrophotometer.

Chemical composition. A knowledge of the chemical composition of clays is helpful in evaluating them for a specific use. How-

Table 3-3

Chemical Analyses of Typical Clays

Constituent	English china clay, washed	N.C. kaolin, washed	Zettlitz kaolin, washed	Georgia sedimentary kaolin	Ball clay, Mayfield, Ky.	Ball clay, Tenn.[1]	Gibbsite, Dutch Guiana	First-grade diaspore, Mo.	Siliceous clay, Rush, Tex.[2]
SiO_2	48.3	46.18	46.87	45.8	56.4	53.96	4.5	10.9	82.45
Al_2O_3	37.6	38.38	38.00	38.5	} 36.0	29.34	58.4	72.4	10.92
Fe_2O_3	0.5	0.57	0.89	0.7		0.98	3.2	1.1	1.08
FeO									
Fe_2S									
MgO		0.42	0.35		tr.	0.30			0.96
CaO	0.1	0.37	tr.	tr.	0.4	0.37	0.4		0.22
Na_2O	} 1.6	0.10	} 1.22	tr.	2.0	0.12			
K_2O		0.58			3.3	0.28			
H_2O									
H_2O+	12.0	13.28	12.70	13.6	7.9	12.82	30.6	13.5	2.40
CO_2									
TiO_2		0.04		1.4		1.64	2.9	3.2	1.00
P_2O_5						.15			
SO_3						.03			
MnO						.02			
ZrO_2		0.08							
Org. C									
Org. H									

Table 3-3 (cont'd.)

Constituent	Brick shale, Mason City, Ia.[2]	Brick clay, Milwaukee, Wis.[2]	Glacial brick clay, Boston, Mass.	Flint fire clay, Cambria, Pa.[3]	Flint fire clay, Carter, Ky.[3]	Semi-flint fire clay, Clearfield, Pa.[3]	Plastic fire clay, Lawrence, O.[3]	Diaspore fire clay, Mo.	Burley flint fire clay, Mo.
SiO_2	54.64	38.07	57.02	44.43	44.78	43.04	58.10	29.2	33.8
Al_2O_3	14.62	9.46	19.15	37.10	35.11	36.49	23.11	53.3	49.4
Fe_2O_3	5.69	2.70	6.70	0.46	1.18	1.37	1.73	1.9	1.9
FeO				0.55	0.74	0.83	0.68		
Fe_2S				0.22	0.14	0.24	0.55		
MgO	2.90	8.50	3.08	0.19	0.55	0.54	1.01		
CaO	5.16	15.84	4.26	0.60	0.77	0.74	0.79		
Na_2O	5.89	2.76	2.83	0.10	0.29	0.46	0.34		
K_2O			2.03	0.55	0.44	1.10	1.90		
H_2O	0.85			0.80	0.84	0.82	2.27		
H_2O+	3.74	2.49	3.45	12.95	13.07	12.44	7.95	12.0	12.0
CO_2	4.80	20.46		0.71	0.07	0.05	0.05		
TiO_2			0.91	1.84	2.22	1.79	1.40	2.7	2.6
P_2O_5				0.21	0.02	0.10	0.17		
SO_3				0.01	0.01	0.01	0.03		
MnO				0.01	0.02	0.01	0.01		
ZrO_2				0.01	0.01	0.01	0.01		
Org. C				0.10	0.11	0.22	0.22		
Org. H						0.03	0.03		

[1]Spinks Clay Company.
[2]Reis, Industrial Minerals and Rocks.
[3]Downs Sheraf, analyst.

ever, this information must be used in combination with the physical properties to obtain a complete picture.

Table 3–3 lists the chemical analyses of a considerable number of typical clays and may be used as a reference when studying new clays.

Unfortunately, only a few analyses are available which are really complete, so we have little idea of the minor constituents in many types of clay. In general, it will be seen that the less pure the clay, the lower will be the amount of combined water. The greater the amount of flux ($CaO + MgO + K_2O + Na_2O + Fe_2O_3$), the lower will be the maturing temperature. The North Carolina residual kaolin is one of the few American clays with the low TiO_2 content of the English China Clays, but the latter contain slightly more feldspar.

Plastic properties

There has never been developed a particularly good quantitative test for plasticity. However, there are certain properties entering into plasticity that are subject to precise measurement, such as the yield point and extensibility, which will be discussed in Chapter 8. We do not have any reliable values for typical clays, and consequently our discussion must be rather general.

The finer-grained clays are highly plastic, but even coarse-grained clays, containing a small portion of montmorillonite, may be quite plastic. On the other hand, shales and flint clays require fine grinding to develop this property. Clays containing appreciable amounts of accessory minerals such as sand lose plasticity.

Dry properties

Drying shrinkage. This property is readily measured by determining either the length or volume change when clay is dried, as discussed in Chapter 12. This property is of great importance when forming large pieces, for a high shrinkage necessitates very slow drying to prevent cracking. In general, the fine-grained, plastic clays have the higher shrinkage values. In Table 3–4 are some typical values of linear drying shrinkage from the plastic to the dried state.

Dried strength. This property is important to facilitate handling ware between the dryer and the kiln. Again the fine-grained clays, especially those containing montmorillonite, are the strongest. Some typical values are shown in Table 3–5.

There has been much speculation about the cause of dry strength, but the most reasonable explanation seems to be the van der Waals forces between the flat faces of the clay particles as they come together, somewhat as shown in Fig. 12–11. It will be noted that the forces are sufficient to orient the plates so that the edges are substantially parallel.

Slaking properties

The time required for a one-inch cube of dry clay to disintegrate after being immersed in water is usually taken as a measure of slaking. This property varies a great deal; flint clays and some shales take an infinitely long time, while washed North Carolina kaolin slakes in 10 minutes. The slaking time has considerable bearing on the process and equipment needed to break down a raw clay to the plastic state.

Firing properties

Since the firing properties, such as shrinkage, porosity, and many other factors, are discussed in Chapters 15 and 16, no reference will be made here to these important characteristics of clay.

Table 3-4

Drying Shrinkage of Clays

Clay	Linear shrinkage in per cent from plastic state
Kaolin, crude	5-8
Kaolin, washed	3-10
Ball clays	5-12
Flint clays (ground)	0.5-6
Sagger clays	3-11
Paving brick clays	1-6
Sewer pipe shales	2-7
Glacial brick clays	3-7

Table 3-5

Dried Strength of Clays

Clay	Modulus of rupture in lb/in^2
Washed kaolin	75-200
Sedimentary kaolin	100-150
Ball clays	150-1200
Glass pot clays	300-1300
Sewer pipe clays	200-600
Sagger clays	100-500
Brick clays	100-1000

REFERENCES

BIRKS, L. S., and FRIEDMAN, H., "Particle Size Determination from X-ray Line Broadening." *J. Appl. Phys.* **17,** 687, 1946.

BLEININGER, A. V., *Properties of American Ball Clays and Their Use in Graphite Crucibles and Glass Pots.* Nat. Bur. Stds., T. P. No. 144, 1920.

DUNN, E. J., JR., "Microscopic Measurements for the Determination of Particle Size of Pigments and Powders." *Ind. Chem. Anal.,* Ed. 2, 59, 1939.

JOHNSON, A. L., and LAWRENCE, W. G., "Surface Area and Its Effect on Exchange Capacity of Kaolinite." *J. Am. Ceram. Soc.* **25,** 344, 1942.

MEYER, W. W., and KLINEFELTER, T. A., *Substitution of Domestic for Imported Clays in Whiteware Bodies.* Nat. Bur. Stds., R. P. 1011, 65, 1937.

NORTON, F. H., and SPEIL, S., "The Measurement of Particle Sizes in Clays." *J. Am. Ceram. Soc.* **21,** 89, 1938.

PARMELEE, C. W., and McVAY, T. N., "An Investigation of Some Ball and China Clays." *J. Am. Ceram. Soc.* **10,** 598, 1927.

SHARRATT, E., and FRANCIS, M., "The Organic Matter of Ball Clays," Part I. *Trans. Brit. Ceram. Soc.* **42,** 111, 1943.

SORTWELL, H. H., *American and English Ball Clays.* Nat. Bur. Stds., T. P. No. 227, 1923

CHAPTER 4

SILICA AND FELDSPAR

Introduction

Silica (SiO₂) is both abundant and widespread in the earth's crust and, in addition, is one of the purest of the abundant minerals. The most common form of silica is quartz, but other forms are found in nature, such as tridymite, cristobalite, vitreous silica, cryptocrystalline forms, hydrated silica, and diatomite. Silica is a very important material for use in ceramics, since it is a major ingredient in glass, glazes, enamels, silica brick, sand-lime brick, abrasives, and whiteware bodies. This extensive use is due to its hardness, relative infusibility, and ability to form glasses.

Feldspars are quite variable in composition but commonly conform to the expression

$$K_x Na_{1-x} \left[\begin{matrix} Al \\ Si_3 \end{matrix} \right] O_8.$$

These minerals are widespread. They are used in ceramics as a fluxing material in whiteware bodies, as a source of alumina in glass, and as a source of alkali in glazes and enamels. It should be remembered that much of the value of feldspars is due to the fact that they are an inexpensive and water-insoluble source of alkalis.

Quartz and other crystalline forms of silica

Quartz. This mineral has been studied very thoroughly and therefore its properties are well known, as shown in Table 4–1. Quartz is one of the few minerals found in large, optically perfect crystals. From these

may be cut specimens for optical instruments and for high frequency electrical oscillators.

The atomic structure of quartz has been shown to consist of a three-dimensional network of SiO₄ tetrahedrons linked into a compact structure, as would be expected from its high specific gravity. The open holes in the structure are so small that other atoms cannot enter and therefore the crystals are always of high purity.

One of the most characteristic properties of quartz is the reversible inversion from the low to the high form at a temperature of 573°C, which will be discussed in Chapter 15.

Quartz occurs as an important constituent of some igneous rocks, such as granite, syenite, and rhyolite, and is a less common constituent of other igneous rocks. It is found in most metamorphic rocks, comprising the major portion of the sandstones and occurring in smaller but definite amounts in clays and shales. Quartz in a pure form is often found in veins running through other rocks.

Crystalline quartz is widespread in nature and many deposits are found in this country. However, good optical cyrstals are not plentiful, and almost all of these have been imported from Brazil. As quartz crystals can now be grown in the laboratory (Fig. 4–1), we are not as dependent on this source as formerly. Rock quartz is not used to any great extent in ceramics because of the cost of grinding and of removing the resultant iron contamination.

The principal source of quartz for the ceramic industry is sandstone consisting of

FIG. 4–1. A quartz crystal grown in the laboratory (Bell Telephone Company).

Table 4-1

Properties of Quartz

Property	Value
Specific gravity	2.651
Hardness (Knoop)	820
Melting point °C	1728
Specific heat (0 - 200°C)	0.203
Coefficient of expansion $\parallel$ (°C)	7.5×10^{-6}
Coefficient of expansion $\perp$ (°C)	13.8×10^{-6}
Index of refraction	1.544
Birefringence	0.009
Crystal system	Hexagonal
Cleavage	Difficult
Dielectric constant	4.5

Table 4-2

Analyses of Glass Sands

Constituent	Mapleton, Pa.	Hancock, W. Va.	Ottawa, Ill.
SiO_2	99.82	99.81	99.61
Al_2O_3	0.12	0.17	0.16
Fe_2O_3	0.017	0.014	0.021
CaO	tr.	0.00	0.050
MgO	tr.	0.00	0.03
Ig. loss	---	---	0.08

lightly bonded quartz grains. For some uses this sandstone is simply disintegrated, but for others it is readily ground in pebble mills with no contamination. In this country we are fortunate in having excellent deposits of these sandstones of high purity. For the glass batch, these sandstones must not only be of high purity, but must also disintegrate readily to a rather uniform grain size. Two types of sandstone fit these conditions, the Oriskany sandstone found from New England to Alabama, and the St. Peter sandstone in Illinois and Missouri. Important deposits are shown in Fig. 4-2. Table 4-2 shows the analysis of some of these deposits, Table 4-3 the chemical specifications for glass sand, and Table 4-4 their natural grain sizes.

Quartzites for silica brick, called gannister, are very firmly consolidated sandstone, so that fracture occurs across the grains and thus permits crushing into fragments of the desired size. These quartzites are found in the Medina formation in Pennsylvania, the Baraboo formation in Wisconsin, and forma-

tions in Alabama and Colorado, as well as a few other places (Fig. 4-3).

Naturally occurring sands are widespread, but seldom are they sufficiently pure for ceramic use. They are employed as foundry sand, as cement aggregate, for sand blasting, for grinding glass, and for many other uses.

Referring only to the ceramic industry, the uses of quartz may be roughly divided as shown in Table 4-5.

Cristobalite. This is another form of silica. It is less pure than quartz, and some geologists feel that these impurities have prevented the formation of the quartz crystal. Cristobalite is not plentiful in nature and is of no use as a source of silica, but as a constituent of fired ceramics it is important and will be discussed in Chapter 15.

FIG. 4-2. Deposits of glass sand in the United States.

Table 4-3

Specifications for Chemical Composition of Glass Sands[1]

Qualities	SiO_2 min.	Al_2O_3 max	Fe_2O_3 max	CaO + MgO max
1st quality (optical glass)	99.8	0.1	0.02	0.1
2nd quality (flint containers and tableware)	98.5	0.5	0.035	0.2
3rd quality (flint glass)	95.0	4.0	0.035	0.5
4th quality (sheet and plate glass)	98.5	0.5	0.06	0.5
5th quality (sheet and plate glass)	95.0	4.0	0.06	0.5
6th quality (green glass containers and window glass)	98.0	0.5	0.3	0.5
7th quality (green glass)	95.0	4.0	0.3	0.5
8th quality (amber glass containers)	98.0	0.5	1.0	0.5
9th quality (amber glass)	95.0	4.0	1.0	0.5

[1]Recommended by the American Ceramic Society and the National Bureau of Standards.

Table 4-4

Screen Analyses of Some Glass Sands

Mesh size	Mapleton, Pa. (per cent retained)	Ottawa, Ill. (per cent retained)	Berkeley Springs, Va. (per cent retained)
14	0	---	---
20	0.5	0	---
28	3.8	3.4	0
35	17.5	30.5	1.1
48	56.9	65.0	33.6
65	90.9	82.1	78.1
100	98.5	92.9	94.6
150	---	98.2	98.6

Tridymite. This form of silica is rare in nature but again is important in fired ware.

Vitreous silica. This glassy material, the result of a lightning flash fusing sandy soil, is found in nature in only a few places. As a manufactured product, however, it is of great value because of its very low coefficient of expansion.

Other forms of silica

Microcrystalline forms. In Europe the silica used in whiteware bodies is flint from calcined flint pebbles found in the chalk beds. The term "potter's flint" has been carried over to this country, even though we use pulverized quartz almost entirely. As far as the final product is concerned, it

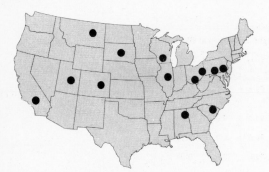

Fig. 4–3. Deposits of gannister in the United States.

Fig. 4–4. Deposits of diatomite in the United States.

Table 4-5

The Use of Quartz in the Ceramic Industry (1947)

Type of use	Per cent of total
Glass sand	74
Gannister for silica brick	22
Whiteware bodies	3
Whiteware glazes	0.5
Enamels	0.5

Table 4-6

Chemical Analysis of Diatomite

Constituent	Lompoc, California	Maryland	Nova Scotia
SiO_2	89.70	79.55	92.78
Al_2O_3	3.72	8.18	2.63
Fe_2O_3	1.09	2.62	1.21
TiO_2	0.10	0.70	---
CaO	0.35	0.25	0.66
MgO	0.65	1.30	0.29
$Na_2O + K_2O$	0.82	1.31	0.46
Ig. Cos.	3.70	5.80	2.22
$SO_3 + Cl$	---	---	---
Total	100.15	99.71	100.25

seems to make little difference which form of silica is used.

The microcrystalline kinds of silica are believed to have formed at relatively low temperatures. There are many types, some of fibrous habit and all very finely crystalline. Most of them contain some water.

Hydrated silica. Opal is typical of this class. This form is an amorphous silica gel with some water. It is of little use in ceramics.

Diatomite. This form of silica consists of the skeletons of diatomes about 10 microns in diameter. The silica is believed to be amorphous. This material is widely found in nearly every bog in the country, but only in a few places does it occur in thick enough layers and of sufficient purity to make it worth while to mine it. Commercial deposits are shown in Fig. 4–4. The best material comes from the west coast. Analyses of some specimens are shown in Table 4–6.

Diatomite is used largely in heat insulators, as the porous skeletons provide myriads of fine pores.

Feldspars

Feldspar minerals. The feldspars used in ceramics are commonly intermediate in composition between the end members, albite $Na\begin{bmatrix} Al \\ Si_3 \end{bmatrix}O_8$, orthoclase $K\begin{bmatrix} Al \\ Si_3 \end{bmatrix}O_8$, and anorthite $Ca\begin{bmatrix} Al_2 \\ Si_2 \end{bmatrix}O_8$. Compositions have been found over much of the area between these end members.

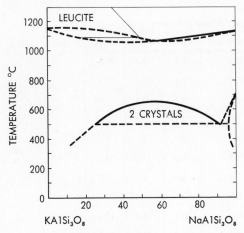

FIG. 4–5. Equilibrium diagram for potash and soda feldspars.

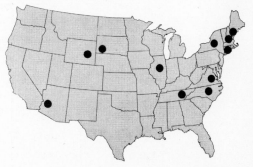

FIG. 4–6. Feldspar deposits in the United States.

The atomic structure of the feldspars is not as yet completely known, but in general they consist of a three-dimensional network of SiO_4 tetrahedrons with a somewhat open structure in which the alkali atoms can rest. To balance the charge a portion of the Si^{++++} is replaced by Al^{+++}. In ceramics the feldspars in the

$$Na\begin{bmatrix} Al \\ Si_3 \end{bmatrix}O_8 - K\begin{bmatrix} Al \\ Si_3 \end{bmatrix}O_8$$

system are of the most importance. This system has recently been worked out by Bowen and Tuttle, and the stability relations are shown in Fig. 4–5.

From the point of view of the ceramist, the crystal form of the feldspar is not important, but the chemical composition is. In addition to the three common feldspars, there occur several others of minor importance in ceramics:

Celsian	$(BaAl_2Si_2O_8)$
Pollucite	$(CsAl_2Si_2O_8)$
Laboratory mineral	$(SrAl_2Si_2O_8)$
Spodumene	$(Li_2Al_2Si_6O_{16})$
Rubidium feldspar	$(RbAl_2Si_2O_8)$

Occurrence of the feldspars. The feldspars are a major constituent of many igneous rocks; for example, granite contains about 60 per cent feldspar. Some of the pegmatites are rich in feldspar, and this is the principal source. Few veins of uniform feldspar are found, and most mining is a hand-sorting operation conducted by farmers living about the milling center. However, the recent development of flotation methods for separation of quartz may change the character of the industry.

The map in Fig. 4–6 shows the location of the important deposits in this country. Sweden is the largest producer of feldspar in Europe, which in general is not well supplied with this mineral.

Properties of commercial feldspars. Table 4–7 shows chemical and mineral analyses of a few feldspars which are representative of those used in the ceramic industry. The high potash spars are used in whiteware bodies and the high soda spars in glasses and glazes. All of the feldspars contain some free quartz, but this is not harmful if it is controlled in amount, as is the case with modern milling methods. However, the iron content must be kept low or color will be introduced into the body or glass.

The specifications for feldspars are given in Commercial Standard CS 23–30. This

Table 4-7

Chemical and Mineral Analyses of Some Feldspars

Constituent	North Carolina	North Carolina	Maine	Canada	Georgia
SiO_2	69.5	70.0	67.8	65.5	65.0
Al_2O_3	17.5	18.1	18.4	18.7	19.5
Fe_2O_3	0.1	0.1	0.1	0.1	0.05
CaO	0.8	1.5	0.3	0.4	0.2
MgO	tr.	tr.	tr.	tr.	tr.
K_2O	8.1	3.5	10.0	12.8	13.1
Na_2O	3.6	6.5	3.0	2.3	2.1
Loss	0.3	0.3	0.3	0.2	0.3
Potash feldspar	47.9	20.7	59.2	75.7	77.5
Soda feldspar	30.6	55.2	25.4	19.5	17.5
Lime feldspar	4.0	7.5	1.5	2.0	1.0
Quartz	14.1	15.7	8.8	1.2	2.5
Other minerals	3.4	0.9	5.1	1.6	1.5

was set up in 1930 by the National Bureau of Standards, and was the first mineral to be definitely specified, a fact very helpful to the feldspar and ceramic industries. Here feldspar is divided into three groups, (1) body spar, (2) glaze spar, and (3) glass spar. In the first group the percentage of silica and the potash-soda ratio are the bases of classification. In the second group soda content is the basis, while in the third group silica content, alumina content, and iron content are specified. In all three groups the particle size is also specified by the amount remaining on certain sizes of screens.

Minerals used in place of feldspar

Nepheline syenite. This mineral is a quartz-free rock consisting largely of nephelite $\left(Na_2 \begin{bmatrix} Al_2 \\ Si_2 \end{bmatrix} O_8 \right)$, soda feldspar and potash feldspar. The only large worked deposit on this continent is at Lakefield, Ontario, where a huge, uniform body is

Table 4-8

Chemical Analyses of Other Fluxing Minerals

Constituent	Nepheline syenite	Aplite	Graphic granite	Cornwall stone
SiO_2	60.2	60.6	72.4	72.6
Al_2O_3	23.7	24.1	14.5	16.4
Fe_2O_3	0.07	0.2	0.3	0.2
CaO	0.4	5.6	0.2	2.1
MgO	0.1	---	---	---
K_2O	5.0	6.3	10.1	4.4
Na_2O	10.0	0.5	2.2	3.1
Loss on ig.	0.5	---	0.3	---
Total	99.9	97.3	100.0	98.8

found. The rock contains specks of magnetite and biotite which are readily removed with a magnetic separator after crushing. This is done at a grinding plant at Rochester, New York.

An analysis of nepheline syenite in Table 4–8 shows that it has a considerably higher content of alumina than does feldspar, and therefore is particularly valuable to the glass industry. However, some is also used in low vitrifying whiteware bodies.

Aplite. This mineral is being produced near Piney River, Va. The rock consists largely of albite, zoisite ($HCa_2Al_3Si_3O_{13}$), and sericite, and smaller amounts of a number of other minerals. A typical analysis in Table 4–8 indicates a high alumina content but also an iron content somewhat higher than that for high-grade feldspars.

Cornwall stone. In England, where feldspars are scarce, this rock, composed of feldspars and quartz, has been largely used in the pottery industry as a flux. The composition varies to some extent, but a typical analysis is shown in Table 4–8.

A similar rock called Carolina Stone has been mined to a limited extent in this country.

Graphic granite. This rock is made up of alternate layers of feldspar and quartz, with an analysis as shown in Table 4–8. It

occurs in large massive deposits of great uniformity and should be an inexpensive source of high silica feldspar. The only thing that holds back its development is the reluctance of manufacturers to use a high silica material, even though they add silica to it in the manufacturing process. On the other hand, should the flotation separation of quartz and feldspar reach a high efficiency, then this rock would serve as an excellent raw material to produce both quartz and feldspar for the ceramic industry.

Production and prices

Crystal quartz. No production figures are available for this material but the consumption in the United States for 1944, the biggest year, was less than a ton. The average retail price is given by McCormick as $7 per pound.

Glass sand. The amount of this material produced in the United States is shown in Fig. 4–7, together with the price. It will

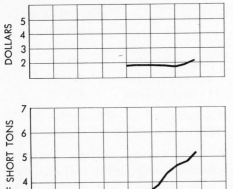

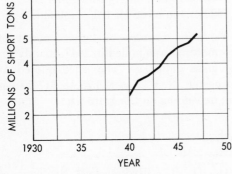

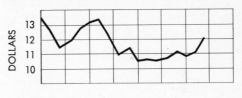

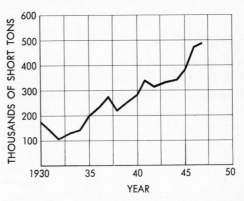

FIG. 4–7. Production and price of glass sand in the United States.

FIG. 4–8. Production and price of ground feldspar in the United States.

Table 4-9

Production of Crude Feldspar in the United States[1]
(in thousands of long tons)

State	1940	1941	1942	1943	1944	1945	1946
Arizona	3.7	5.0	6.0	3.0	9.8	8.8	8.7
California	2.7	4.5	5.1	2.0	1.2	0.8	---
Colorado	34.1	42.3	33.6	20.7	15.8	26.8	37.3
Connecticut	24.4	13.7	12.8	11.6	11.4	11.7	16.6
Georgia	---	---	---	---	---	---	0.1
Maine	18.4	22.6	8.8	6.7	8.0	11.0	18.9
New Hampshire	38.6	52.2	46.5	44.0	42.6	48.3	63.6
New York	5.0	5.0	4.0	4.0	4.0	3.0	4.8
North Carolina	79.3	100.0	93.6	112.1	122.8	148.5	230.4
Pennsylvania	0.3	0.4	0.3	0.4	0.5	0.2	---
South Dakota	54.7	59.0	56.5	70.9	64.8	68.4	74.5
Texas	---	0.1	0.1	0.1	---	---	0.2
Virginia	21.7	22.0	24.3	20.6	24.0	29.1	33.0
Wyoming	7.8	11.8	14.0	18.0	22.4	17.0	20.3

[1] Figures from Industrial Minerals and Rocks, p. 350.

be seen that this is one of the least expensive of our ceramic raw materials.

Potter's flint. The amount of flint produced by the mills in this country is not known exactly but is less in tonnage than feldspar. The price runs from $10 to $15 per ton.

Diatomite. The production of diatomite in the United States was 215,000 short tons in the year 1946. California was the principal producer. The average price was $20 per ton in that year.

Feldspar. The production of crude feldspar is shown in Table 4-9. North Carolina is by far the largest producer, almost equaling the total of all the other states at the present time. The amount of ground feldspar sold and its average price is shown in Fig. 4-8. In 1948 the ton price of ground feldspar was:

(Depending on
grade and
sizing)

Glass feldspar	$11–14
Enamel feldspar	$14–16
Pottery feldspar	$18–21

Carload lots f.o.b. Spruce Pine, N. C. $2 per ton more for bag shipment.

Nepheline syenite. A large amount of this material was milled in 1947, mainly for the glass industry. The price quoted at Rochester, N. Y. was $13.75 to $17.25 per ton, depending on the grain size.

REFERENCES

BOWEN, N. L., and TUTTLE, O. F., "The System NaAlSi₃O₈-KAlSi₃O₈-H₂O." *J. Geol.* **58**, 489, 1950.

HALE, D. R., "Hydrothermal Synthesis of Quartz Crystals." *Ceramic Age*, November, 1950.

Industrial Minerals and Rocks, 2nd ed. A.I.M.E., 1949.

KNIGHT, F. P., JR., "Commercial Feldspars Produced in the United States." *J. Am. Ceram. Soc.* **13,** 532, 1930.

LADOO, R. B., *Non-metallic Minerals.* McGraw-Hill Book Co., Inc., New York, 1925.

Minerals Year Book. U. S. Bur. of Mines, Washington, D. C., 1947–1948.

SOSMAN, R. B., *The Properties of Silica.* Chemical Catalog Co., Inc., New York, 1927.

CHAPTER 5

MAGNESITE, LIME, DOLOMITE, AND CHROMITE

Introduction

Because of their high softening point and their stability in contact with many slags, the basic materials described in this chapter are of particular interest to the manufacturer of refractories. However, lime and magnesia are used in some whiteware bodies, glazes, enamels, and, quite extensively, in glass. Chromium compounds serve as an important source of color in ceramics.

Magnesite

This mineral is the normal carbonate, $MgCO_3$. Magnesite is often associated with calcite, $CaCO_3$, to form the common basic rock, dolomite. However, it is a mixture of crystals rather than a solid solution. On the other hand, $MgCO_3$ does form a continuous isomorphous series with siderite, $FeCO_3$.

Origin. Magnesite may occur in two general forms; the crystalline, and the dense or cryptocrystalline. The crystalline variety is usually formed in nature by the alteration of dolomite through the percolation of magnesium solutions. The dense variety is often an alteration product of serpentine types of rock.

Occurrence. In this country crystalline magnesite is being regularly mined only in the state of Washington. A deposit in Quebec, somewhat higher in lime, has been mined, and mining has been carried on both in Nevada and California. The largest and most important deposits are those in Styria, Austria, which in the past have contributed

much of the world supply for metallurgical uses. This material is high in iron oxide, a desirable characteristic for this use. The U. S. S. R. has excellent deposits of magnesite and has recently become the world's largest producer. In Table 5–1 are given some typical analyses of magnesites. Other deposits are shown on the map of Fig. 5–1.

Before the Second World War magnesia was being produced on a small scale by recovery from sea water. Since then this production has been greatly increased, so that a considerable tonnage used in refractories now comes from this source. The general method is discussed in Chapter 7. Plants for this operation are located in California, Texas, Maryland, and New Jersey. Some magnesia is also made from bitterns (taken out of salt wells).

There is also the possibility of using modern froth flotation methods to concentrate contaminated deposits of magnesite for producing a low lime, low silica product. This possibility is an attractive one, as deposits close to the steel industry in the eastern part of the country would then become available sources.

Burning. Before shipping, most magnesite is calcined to drive off the CO_2. Two grades are generally made: caustic magnesite calcined to 700–1200°C, leaving 2 to 7 per cent of CO_2 in the rock, and dead-burned magnesia fired at 1450–1500°C, which removes practically all the CO_2 and sinters the grain into a dense, stable mass. The caustic magnesia is largely used in cements of the

Table 5-1

Chemical Analyses of Dead-Burned Magnesites[1]

Constituent	Austrian	Austrian	Manchukuo	Washington	Greece	California sea water
SiO_2	5.8	1.0	3.7	4.9	6.6	5.1
Al_2O_3	1.7	1.0	1.0	1.5	4.4	0.5
Fe_2O_3	4.0	6.9	1.5	3.4		0.2
CaO	5.0	2.1	1.6	2.8	2.4	1.8
MgO	83.0	88.6	92.0	87.1	86.4	91.7
Ig. loss	0.2	0.3	0.1	0.1	0.2	0.3
Total	99.7	99.9	99.9	99.8	100.0	99.6

[1] From Refractories.

Sorel type, while the dead-burned material is used in the refractories industry as grain or in the form of bricks. The calcination was formerly carried out in vertical shaft kilns, but more recently the rotary kiln has been found to give a more uniform product. For refractory use the Washington and sea-water magnesite often have iron ore or mill scale added in the burning process for better sintering.

Uses. The principal use of magnesite is for refractories. For this purpose it is indispensable, since it has one of the highest fusion points known (2800°C) and is resistant to many types of slag.

The ceramic industry requires dead-burned magnesite grain for open hearth bottoms and other plastic uses. A large amount is used in fired magnesite brick and for chemically bonded brick. Purer grades of the carbonate are used for low temperature insulation, in whiteware bodies, in glazes and enamels, and as an ingredient in glass batches. The use of very pure MgO as a super-refractory may have future possibilities.

Lime

This is the carbonate of calcium, $CaCO_3$. The mineral is generally calcite, a trigonal crystal of diverse habit, but there is an orthorhombic form, aragonite, found especially in shells. The natural limestone is used in ceramics in ground form, in glazes, enamels, and glasses. When more purity is required, a chemically precipitated product known as "whiting" is employed. However, the bulk of the lime produced is used in agriculture and in plasters.

The limestone, much like magnesia, is burned in kilns to produce quicklime or CaO. The latter is unstable when exposed to air, as it both hydrates and carbonates readily. Dry lime hydrate is now used largely in plasters and mortars. Lime may be fused and

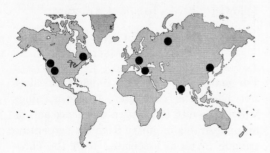

FIG. 5-1. Deposits of magnesite.

Table 5-2

Chemical Analyses of Limestones and Dolomites[1]

Constituent	Rockland, Me.	Bellefonte, Pa.	Knoxville, Tenn.	Hollow, Pa.	Marblehead, Ohio	Manitowoc, Wis.
SiO_2	1.3	1.4	0.2	0.8	1.8	0.2
Al_2O_3	0.2	0.3	0.1	0.6	0.4	0.2
Fe_2O_3	0.4	0.4	0.1	0.5	0.1	0.1
$CaCO_3$	95.9	96.4	98.9	54.7	87.7	54.8
$MgCO_3$	2.3	1.6	0.8	43.7	10.1	45.0
Total	99.9	100.1	100.1	100.3	100.1	100.3

[1] From Industrial Minerals and Rocks.

Table 5-3

Chemical Analyses of Chrome Ores

Constituent	Turkey	Rhodesia	Phillippines	New Caledonia	Russia	Cuba
Cr_2O_3	46.6	45.4	32.1	54.5	46.2	30.5
SiO_2	6.7	7.5	5.3	3.1	4.0	6.1
Al_2O_3	12.5	13.8	27.6	11.0	14.6	27.5
FeO	12.9	15.1	13.0	19.5	15.6	14.2
CaO	1.2	0.5	1.1	1.5	0.3	0.9
MgO	17.3	13.6	18.2	8.0	15.4	18.3
Total	97.2	95.9	97.3	97.6	96.1	97.5

crystallized, in which condition it is more stable. However, lime refractories have never been used to any great extent.

Limestone is widespread, occurring in practically every state in the union, but only a few deposits are especially pure. Table 5-2 gives the chemical analyses of some limestones.

Dolomite

This rock is a mechanical mixture of magnesite and calcite crystals of varying proportions. It is used in ceramics as a source of lime and magnesite for whiteware bodies and for glass. Great quantities are also used in the dead-burned form for the bottoms of open-hearth furnaces.

This mineral is of widespread occurrence and variable composition. Chemical analyses are given in Table 5-2.

Chromite

Mineral form. This mineral is a spinel Cr_2FeO_4. Actually much of the Cr_2O_3 and FeO may be replaced in solid solution as follows to leave a balanced structure:

$$\left.\begin{array}{c} Cr_2O_3 \\ Al_2O_3 \end{array}\right\} \quad \left\{\begin{array}{c} FeO \\ MgO \end{array}\right.$$

For many purposes, this replacement does not harm the chromite. However, silica will not go into this lattice and hence acts as an undesirable impurity from the serpentine gangue material associated with the chromite.

Table 5-4

Production and Prices in 1945

	U. S. production (short tons)	World production (short tons)	Average price (dollars/short ton)
Magnesite (dead-burned)	336,000	--------	6.93
Lime	5,921,000	--------	7.76
Dolomite (dead-burned)	1,187,000	--------	8.95
Chromite	15,000	2,000,000[1]	34.60

[1]Estimated.

Origin. Chromite ores are believed to originate both from magmatic crystallization and by precipitation from hydrothermal solutions. Mining is done by the usual methods, with hand sorting to remove the gangue. Concentration methods such as flotation have not as yet been used to any extent, but magnesia is sometimes added to form forsterite with the free silica present.

Unfortunately, the United States has very limited sources of chrome ore, and must therefore import nearly all of the ore used here. Chrome ore is mined in Turkey, Rhodesia, New Caledonia, the Philippines, Cuba, the U. S. S. R., and a few other places, as shown on the map in Fig. 5–2. Typical analyses are given in Table 5–3.

Uses. Chromite is used in ceramics largely as a refractory in the form of burned bricks, chemically bonded bricks, and as plastics. For this purpose a low silica ma-

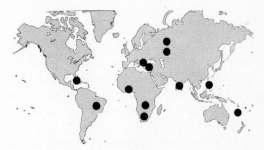

Fig. 5–2. Deposits of chrome ore.

terial is desired. Chromite is also a source of chromic oxide and other compounds used as a color or stain.

Production and prices

The figures in Table 5–4 give an idea of the production and prices of the materials covered in this chapter. For more complete data, refer to *Industrial Minerals and Rocks* and the *Mineral Year Book*.

REFERENCES

BIRCH, R. E., and WICKEN, O. M., "Magnesite and Related Minerals," *Industrial Minerals and Rocks*, 2nd ed. A. I. M. E., 1949, pp. 521–541.

COLBY, S. F., *Occurrences and Uses of Dolomite in the United States*. U. S. Bur. of Mines, Inf. Cir., 1941.

DODD, A. E., and GREEN, A. T., "The Genesis of Chrome Ores." *Iron and Steel Inst.*, Spec. Rep. 32, p. 43, 1946.

HUGELL, W., and GREEN, A. T., "Constitution of Chrome Ores." *Iron and Steel Inst.*, Spec. Rep. 32, p. 31, 1946.

JOHNSTON, W. D., JR., and THAYER, T. P., "Chromite," *Industrial Minerals and Rocks*, 2nd ed. A. I. M. E., 1949, pp. 194–206.

STEDMAN, G. E., "Magnesium from Sea Water." *Metals and Alloys* 20, 941, 1944.

CHAPTER 6

OTHER NONCLAY MINERALS

Introduction

The minerals described in this chapter are all used in ceramics, but not in major amounts.

Fluxes

Lithium minerals. The important lithium minerals are spodumene ($Li_2Al_2Si_4O_{12}$), lepidolite ($LiKAl_2F_2Si_3O_9$), amblygonite ($Li_2F_2Al_2P_2O_8$), and petalite ($LiAlSi_4O_{10}$). The alkali content of these minerals is shown in Table 6-1.

In a few cases spodumene is used in the mineral form as a source of lithium, but in most cases the lithium salt is extracted and used in the pure form in glasses, glazes, and especially in acid-resisting enamels. Some lithium salts are extracted from brines. Not more than a small fraction of the output of lithium is now used in the ceramic industry, but it is expected that this use will increase.

Barium minerals. Barium compounds are used in various branches of ceramics. They act as fluxing compounds in glazes, glasses, and enamels, and in heavy clay products they form insoluble barium sulphate to prevent scumming.

The raw materials commonly used are barite ($BaSO_4$) and witherite ($BaCO_3$). The former is found in this country in a number of places, particularly Missouri, Georgia, Tennessee, and California. The pure carbonate is made chemically and is used in this form for most ceramic purposes. There is no information on the amount of barium compounds used in the ceramic industry, but the tonnage is not large.

Table 6-1

Alkali Content of Some Lithium Minerals

Mineral	Li_2O	Cs_2O	Rb_2O
Spodumene	7%		
Lepidolite	4%	0.8%	3.0%
Amblygonite	9%		
Petalite	7%		

Fluorine minerals. For ceramic use, the most important mineral containing fluorine is fluorite (CaF_2), occurring in the rock, fluorspar. It is concentrated from natural deposits to give a product of 90–98 per cent CaF_2, the principal impurity being silica. There are many other fluorine-containing minerals, but none of them is used to a great extent in ceramics.

Fluorspar is used in many optical glasses of low index of refraction and in enamels. It is found in a number of workable deposits, the principal ones in this country being in Illinois, Kentucky, and some of the western states. About 340,000 short tons of fluorspar were produced in this country in the year 1947, but only a small portion of this was used in ceramics, the remainder going mainly into metallurgical slags.

Phosphate minerals. The common phosphorous-bearing mineral is apatite [$Ca_5(Cl,F)(PO_4)_3$] and is found in many deposits in Florida, Tennessee, Utah, Wyoming, Idaho, and Montana. Its main use, of course, is for agriculture, but a few per cent of the total is used in ceramics for glasses and enamels. The whiteware body, bone china, derives its phosphates from calcined bones.

Magnesium silicates

Talc. This mineral is of widespread occurrence and variable composition. The pure mineral may be expressed as $(OH)_2Mg_3Si_4O_{10}$. It is a layer-structure mineral analogous to montmorillonite (Fig. 1–6), except that the cations in the octahedral layer amount to $6Mg^{++}$ rather than $3Al^{+++} + 1Mg^{++}$. For this reason the mineral is soft and cleaves readily, but does not have an expanding lattice. The magnesium is often partially replaced by calcium and iron. Many other minerals, such as magnesite, tremolite, and serpentine, are often associated with talc in the industrial product.

Talc, formed as a secondary product from basic rocks, is found in many places in this country, as shown by the map in Fig. 6–1. Other deposits of importance are in Manchuria, France, India, Canada, and Norway.

For ceramic use a pure form of talc is desired. Analyses of some ceramic talcs are shown in Table 6–2.

Talc is used largely in electrical insulator bodies of the steatite type, where a magnesium silicate composition having good me-

FIG. 6–1. Talc deposits in the United States.

chanical and electrical properties is required. It is also used in some whiteware bodies, particularly tiles and ovenware.

The production and prices of talc are shown in Table 6–3, but it is hard to say what proportion is used in the ceramic industry.

Asbestos. This unique rock may be readily broken down into a soft fibrous structure, with fibers varying in length up to three inches or even more. There are two main types, the most important of which is chrysotile, a variety of serpentine. The other is an amphibole, which is less ther-

Table 6-2

Chemical Analyses of Talcs and Pyrophyllite

Constituent	New York[1]	Vermont[1]	North Carolina[1]	California[1]	Georgia[1]	Pyrophyllite[2]
SiO_2	66.2	56.3	61.4	59.6	41.0	63.5
Al_2O_3	1.1	3.2	4.4	1.7	4.2	28.7
Fe_2O_3	0.6	5.4	1.7	0.9	5.9	0.8
MgO	25.7	27.9	26.0	30.0	28.6	tr.
CaO	2.3	0.4	0.8	0.8	4.8	tr.
MnO	0.2	0.1	---	---	---	---
$K_2O + Na_2O$	---	0.9	---	0.3	---	0.4
CO_2	0.6	0.4	---	---	---	---
Comb. water	3.9	5.7	5.1	5.9	15.5	5.9

[1]From Industrial Minerals and Rocks.
[2]From Ec. Pap. No. 3, N.C. Geol. Surv.

Table 6-3

Production and Prices of Talc in 1946[1]

Source	Short tons	Average value in dollars per short ton
California	78,000	18
New York	112,000	19
Vermont	75,000	11
U. S. total	407,000	14
World production	1,200,000	--
Imports	18,000	21

[1]From Industrial Minerals and Rocks.

mally stable. The important minerals in asbestos rock are tremolite [$H_2Ca_2Mg_5$-$(SiO_3)_8$], anthophyllite [$H_2Mg_7(SiO_3)_8$], and chrysotilite [$H_4Mg_3Si_2O_9$]. These all have the double chain structure of the metasilicates, as shown in Table 1-4. This accounts for their ready cleavage into fibers, as the bonding between the chains is weak.

The most important deposit on this continent is in the Province of Quebec, and ex-

tensive deposits also occur in the U. S. S. R. and in Africa.

Asbestos is used in various types of low- and medium-temperature insulation and in fireproof materials, since the fibers considerably increase the toughness and flexibility of a product.

Aluminous minerals

Corundum (Al_2O_3). In the impure form known as emery, this mineral was used since early times as an abrasive, but it has now been replaced almost entirely by fused alumina. A small amount of massive corundum is imported from South Africa for use with fused alumina in some types of grinding wheels.

Sillimanite minerals. These minerals, with the theoretical composition of Al_2SiO_5, are of interest as a source of alumina in refractories. They include sillimanite, kyanite, and andalusite, the properties of which are shown in Table 6-4.

Table 6-4

Properties of the Sillimanite Minerals

Mineral	Physical properties			
	Crystal system	Cleavage	Hardness.	Specific gravity
Andalusite	Orthorhombic	(110) distinct	7.5	3.13-3.20
Kyanite	Triclinic	(100) perfect (010) good	4 ‖ length 7⊥ length	3.53-3.67
Sillimanite	Orthorhombic	(010) perfect	6-7	3.23-3.24

Mineral	Optical properties		Thermal properties	
	Refractive index	Optical characteristics	Decomposition temp.	Volume change
Andalusite	1.63 1.64 1.64	Biaxial negative	1350-1450°C	Very slight increase
Kyanite	1.71 1.72 1.73	Biaxial positive	1100-1480°C	Marked increase
Sillimanite	1.66 1.66 1.68	Biaxial negative	1550-1650°C	Slight increase

About the only known commercial deposit of andalusite in this country is in the Ingo Mountains of California. The ore occurs at altitudes of 7,000 to 9,000 ft in lenses bordered by sericite schists. The only other good source of this mineral is in the U. S. S. R.

Kyanite is found in the United States in the Piedmont belt of schists extending from Virginia to Alabama. The most important deposit is in the Celo Mountains in North Carolina. However, in this country the product is of fine grain size and thus is not as suitable for refractory use as the massive material imported from India.

Sillimanite has not been used as extensively as the other minerals, but a number of deposits are known both in this country and abroad, and those may be worked when efficient concentrating methods are developed. The domestic production of sillimanite minerals was about 15,000 short tons in the year 1949, and about an equal amount was imported. The price in the same year was $50 per long ton of crude ore and $100 for calcined and ground ore.

Topaz [$Al_2SiO_4(F,OH)$]. This mineral has been used in refractories to a small extent. Deposits are found in South Carolina.

Dumortierite ($HBAl_8Si_4O_{20}$). This has been considered as a constituent of whiteware bodies. One workable deposit is known, in Nevada.

Pyrophyllite [$(OH)_2Al_2Si_4O_{10}$]. This is a mineral having a structure similar to montmorillonite except that it is perfectly crystalline and has no replacement of Si^{++++} with Al^{+++}. The most important deposit of this mineral is in North Carolina, and a typical analysis of it is shown in Table 6–2. This mineral has been used to some extent in whiteware bodies, but as yet is not an important ceramic material.

Glaze and glass minerals

Lead compounds. In modern ceramics lead salts are chemically prepared and thus are of high purity. The ones commonly employed are red lead (Pb_3O_4) and white lead [$2PbCO_3 \cdot Pb(OH)_2$]. These are used in glasses, glazes, and enamels as a basic flux. From the point of view of health, the insoluble $PbSi_2O_5$ is desirable where it can be used.

Zinc compounds. Zinc oxide (ZnO) and zinc carbonate ($ZnCO_3$), chemically prepared, are used in glazes — particularly Bristol glazes.

Boron compounds. Boron in the form of boric acid (H_3BO_3) or borax ($Na_2B_4O_7 \cdot 10H_2$) is used in glasses, enamels, and fritted glazes. Boron acts both as a glass former and as a flux. The deposits in the deserts of Nevada and California are the most important sources of the raw material, but commercial deposits occur in many other parts of the world. The mineral colemanite ($2CaO \cdot 3B_2O_3 \cdot 5H_2O$), being insoluble in water, is sometimes used in raw glazes.

Tin oxide (SnO_2). This chemical has been used as an opacifying agent in glazes, but because of its high cost has largely been replaced by zirconia and titanium compounds. There is no source of tin within the United States.

Refractory minerals

Zirconium minerals. The raw materials are found mainly as the oxide in the form of zirkite or baddeleyite (ZrO_2) or as zircon ($ZrSiO_4$). Deposits are shown on the map of Fig. 6–2.

Most of the zirconia-containing materials used in ceramics are chemically extracted from the ore and used as opacifiers in glazes and enamels, or as refractories, after fusion with a few per cent of lime that acts as a

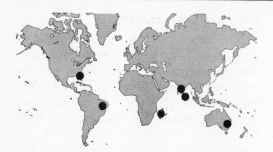

Fig. 6–2. World deposits of zirconia minerals.

Fig. 6–3. World deposits of uranium minerals.

stabilizer to prevent crystal inversion. All commercial zirconia contains about 4 per cent hafnia.

Beryllium minerals. The chief source of supply is the mineral beryl [$Be_3Al_2(SiO_3)_6$], often associated with pegmatites. There are few deposits of the mineral extensive enough to justify mining it alone, but often it can be taken out with feldspar. Beryl, containing about 12 per cent beryllia, is chemically processed to give a pure product for ceramic use. Beryllia is used in some glasses, but its main use is in refractories because of its remarkable nuclear properties and its extraordinarily high thermal conductivity. It should always be kept in mind that beryllium compounds, especially the oxide, are among the most toxic materials known. They should never be handled without proper safeguards.

Titanium minerals. Rutile (TiO_2) and ilmenite ($FeTiO_3 \cdot FeFeO_3$) are common minerals but are seldom concentrated in a natural deposit except in the case of some beach sands. The important rock deposits are in New York State, Norway, Quebec, Virginia, and North Carolina. In addition, there are many beach sand deposits. Natural rutile is used as a color in bodies and glazes. Most of that used in ceramics is chemically purified. It is used in enamels and glazes as an opacifier, and in electrical porcelain

bodies of high dielectric constant. For the latter purpose the titanates of the alkaline earth metals are also used.

Thorium minerals. The monazite sands are the chief source of thoria. Some of the important deposits are in Florida, India, Brazil, and Australia. Besides thoria, they contain scandia, yttria, and lanthana, as well as the rare earth oxides. Thoria is used for crucibles and other special refractories because of its extremely high melting point, reported as 3050°C. It is also a fissionable material and thus of restricted use. It should be noted that thorium compounds slowly generate a radioactive gas in storage, so any containers of this material should be opened out of doors or under a hood.

Cerium minerals. This element is also found in monazite sands. It has been used in some special glasses, as a polishing medium for optical glass, and as a special refractory.

Uranium oxide. The uranium compounds at one time were an important red colorant in glazes, but now such use is restricted because of the fissionable nature of this element. At present, much intensive prospecting is going forward to find new sources of uranium, so that fields are developing constantly. Some of the principal deposits are shown on the map in Fig. 6–3.

Rarer elements. The rare earth elements,

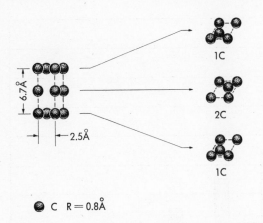

FIG. 6–4. The unit cell of graphite showing layer structure.

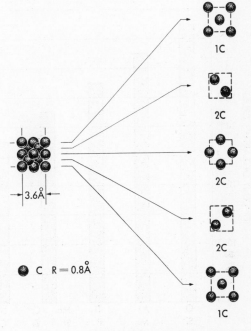

FIG. 6–5. The unit cell of diamond.

as well as yttria and lanthana, have been used in special glasses and as special refractories. At present the costs are too high for extensive use.

Carbonaceous materials

Graphite. This is a black, flaky mineral consisting entirely of carbon. The atomic structure (Fig. 6–4) consists of widely spaced layers of hexagonally packed carbon atoms, which accounts for its softness and flakelike character. It is found in nature in a number of places such as Ceylon, Madagascar, and Korea. It may also be formed from coke in the electric furnace. Graphite is of great value as a refractory because of its high electrical and thermal conductivity and its high stability in a reducing atmosphere. Mixed with clay, graphite makes excellent crucibles for metal melting. It also has nuclear properties that make it valuable in a reactor.

Coal and coke. Coal is a widely distributed fuel that may be changed to coke by distilling off the volatile matter. The coke may be finely divided and pressed into a solid mass by means of an organic binder

which is later carbonized by heating in a reducing atmosphere. Carbon is used for electrodes, crucibles, refractory blocks, and many other purposes involving high temperature service.

Diamond. This extraordinary mineral is a form of carbon with a hardness far above

Table 6-5

Hardness Values for Some Minerals[1]

Material	Knoop hardness
Diamond	8000
Boron carbide	2800
Silicon carbide	2500
Corundum	2000
Topaz	1300
Quartz	800

[1]Thibault, N. W., and Nyquist, H. L., Trans. Am. Soc. of Metals, 271, 1946.

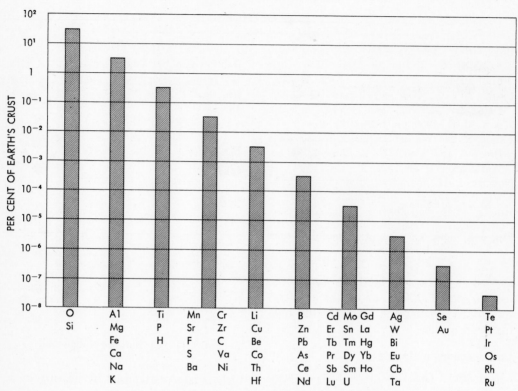

FIG. 6–6. Abundance of elements in the earth's crust.

that of any other known material, as shown in Table 6–5.

The structure of the diamond is based on a cubic unit cell with eight atoms (Fig. 6–5). This structure, while characteristic of hard, brittle materials, does not completely explain the extreme hardness of this mineral. The diamond is an efficient abrasive, either as a powder or bonded into a wheel. Only its scarcity prevents it from taking over a large share of the abrasive industry.

Relative abundance of the elements

To summarize, the availability of ceramic materials depends on the degree to which nature has concentrated them and on the accessibility of the deposits. The abundance in the earth's crust (see Fig. 6–6) is not the important factor, for many such commonly used elements as mercury, silver, and antimony are less abundant than many of the so-called rarer elements.

REFERENCES

Bowles, O., *Asbestos — Domestic and Foreign Deposits*. U. S. Bur. of Mines, Ind. Circ. 6790, 1934.

Engel, A. E. J., "Talc and Ground Soapstone," *Industrial Minerals and Rocks*, 2nd ed. A. I. M. E., 1950, pp. 1018–1041.

Ladoo, R. B., *Talc and Soapstone, Their Milling Products and Uses*. U. S. Bur. of Mines, Washington, D. C., 1923.

Norton, F. H., *Refractories*, 3rd ed. McGraw-Hill Book Co., Inc., New York, 1949, Chap. 4.

Riddle, F. H., and Foster, W. R., "The Sillimanite Group," *Industrial Minerals and Rocks*, 2nd ed. A. I. M. E., 1950, pp. 893–926.

Roy, R., Roy, D. M., and Osborn, E. F., "Composition and Stability Relationships Among the Lithium Aluminates: Encyptite, Spodumene and Petalite." *J. Am. Ceram. Soc.* **33**, 152, 1950.

Thoenen, J. R., and Burchard, E. F., *Bauxite Resources of the United States*. U. S. Bur. of Mines, Repts. Invest. 3598, 1941.

Thurnauer, H., "Review of Ceramic Materials for High-frequency Insulation." *J. Am. Ceram. Soc.* **20**, 368, 1937.

CHAPTER 7

MINING AND TREATMENT OF THE RAW MATERIALS

Introduction

The mining and preliminary treatment of the raw materials are important steps in the production of ceramic ware, and therefore warrant a brief treatment here.

In the past, the mining of clays was often done on such a small tonnage basis that the methods employed were wasteful of labor compared with those used in the mining of metals. However, many mines are now using such efficient methods that the cost of a ton of clay today is little more than it was a generation ago, in spite of a great increase in wage rates.

Great strides have also been made in improving the methods of treatment to give the customer a pure and uniform product. Examples of this are washed kaolins and ground feldspar. Also, as supplies of the higher grade materials are exhausted, more effort must be given to purification of lower grade deposits.

Mining

Open pit methods. The majority of clays are mined in open pits by stripping the overburden and taking out the useful clay. The stripping may be done by power shovels working on a face or, in some mines, by bulldozers and power scrapers of large capacity. The maximum economical depth of overburden depends on both the thickness and the value of the underlying clay layer.

The clay itself may be removed by power shovels, shale planers, or, if it is hard, by blasting. It is now commonly transported to the plant by truck rather than by rail. In some cases, such as in Cornwall, the clay is washed down into a sump by a powerful stream of water. Figure 7–1 shows sections of a number of typical clay mining operations.

Harder materials like quartz rock, feldspar, and kyanite are quarried by the ordinary methods of drilling and blasting. Glass sands are often removed hydraulically.

Underground methods. Coal measure clays are usually taken from deep strata, using the conventional methods of sinking a shaft and then tunneling out into the proper level. As in all mining, a careful plan should be made, based on diamond drilling. Good timbering and clean, level tracks are essential for efficient operation.

Methods of comminution

Theory. The reduction of a particle to two or more parts may take place in several ways, as shown in Fig. 7–2. In (a) the fracture takes place under simple compression if the material is brittle, in (b) by compression impact. In (c) the break is made by an impact too low in energy to fracture the whole piece but sufficient to remove a small corner. Breakage may occur as a result of one particle striking another at high velocity, as in (d). In some cases subdivision occurs by abrasion (e). In the last case (f), a shredding action takes place in soft materials when a cutting tooth actually shaves off a fragment. In actual practice it is impossible to divide comminuting machines into rigid groups like those above, for in

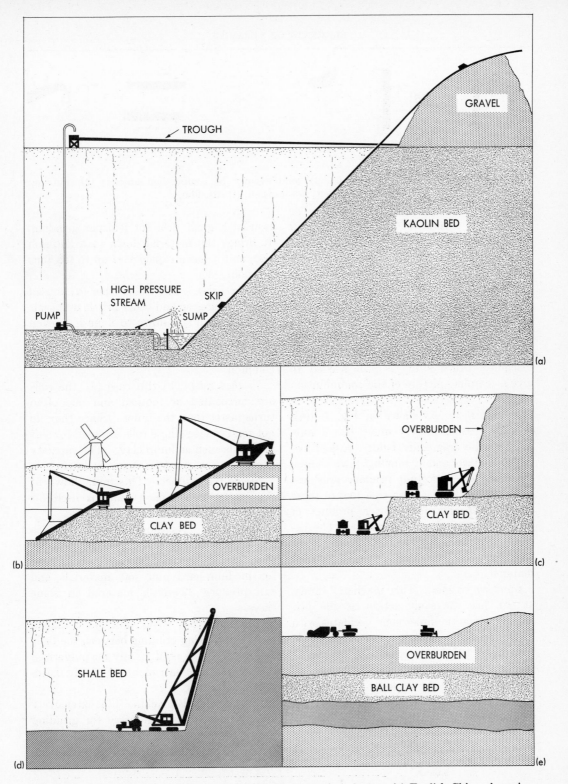

FIG. 7–1. Some typical mining operations in the ceramic industry: (a) English China clay mine in Cornwall; (b) removing fire clay in Germany; (c) removing Georgia kaolin with power shovels; (d) taking out a hard clay with a shale planer; (e) stripping ball clay with earth-moving equipment.

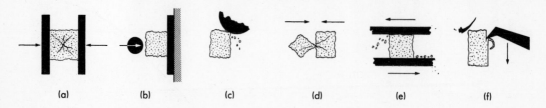

FIG. 7–2. Principles of comminution: (a) compression; (b) compression impact; (c) nibbling; (d) self-impact; (e) abrasion; (f) shredding.

most cases several of the actions occur simultaneously.

In general, the energy required to break down a piece of material is proportional to the new surface area produced; thus the time and power expended increase rapidly as the size is reduced. The shape of the particles, as well as their distribution, varies both with the material and with the type of machine. Much study is still needed to give a complete picture of the comminution process.

Jaw crusher. As shown in Fig. 7–3(a), the jaw crusher is very much like a nut-cracker, the fragments being crushed between a fixed and a moving jaw. These machines are important primary crushers, taking a feed of pieces up to a foot or more in diameter and delivering a product of pieces 1 to 3 inches in diameter. The capacity may be as high as 100 tons per hour. The feed may be hard or medium-hard material.

Gyratory crusher. This machine, shown at (b), has the same action as the jaw crusher, since the inner cone oscillates with a circular motion but does not revolve. It is used for brittle materials like magnesite and limestone, and has a capacity of up to several hundred tons per hour.

Cone crusher. This machine, shown at (c), operates on the same principle as the previous one, except that the cone is more obtuse to give a greater discharge area. It

is used a great deal for feldspar grinding. It brings the material down to a 20-mesh size, and it has a capacity of up to ten tons per hour.

Roll crusher. The roll crusher (d) is used for crushing grog and other brittle materials from one-inch size down to 8 mesh or even finer. The squeezing action is continuous, and capacities up to 10 tons per hour are common.

Toothed rolls. In this case (e), the rolls are corrugated or toothed and one often turns faster than the other. Since they do not tend to clog, these rolls are used for soft materials such as lump clay. The capacity is great, as much as 180 tons per hour.

Single roll crusher. This mill, shown at (f), has a high capacity and is used for medium soft materials like limestone.

Roller mill. As shown at (g), this is a roll crusher in which the grinding pressure is exerted by centrifugal force. It is used for medium-hard and soft materials and can produce 325-mesh material in some cases.

Ball pulverizer. This mill (j) is like a huge ball bearing grinding between the balls and races. It is used mainly for pulverized coal, but is also satisfactory for soft materials like bauxite.

Dry pan. This machine (h) is extensively used in the ceramic industry for grinding gannister, grog, shale, or flint clay. There are many modifications, but in general the

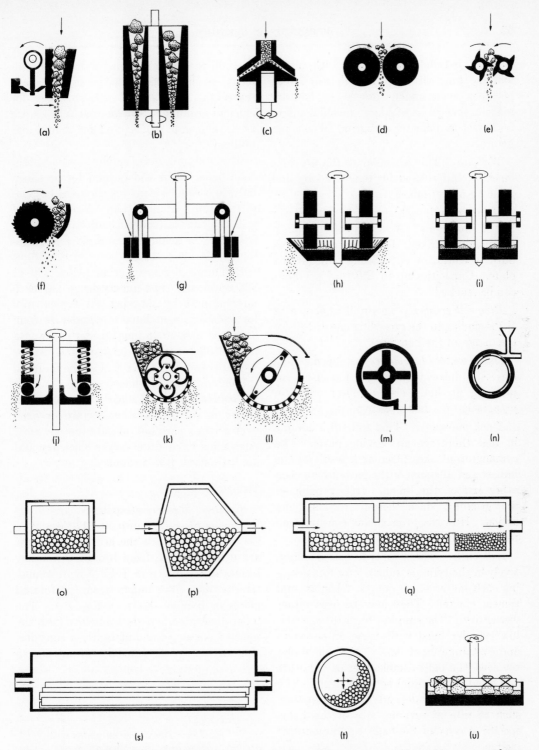

FIG. 7–3. Comminuting equipment: (a) Jaw crusher; (b) gyratory crusher; (c) cone crusher; (d) smooth roll crusher; (e) toothed-roll crusher; (f) single-roll crusher; (g) roller mill; (h) dry pan; (i) wet pan; (j) B and W Mill; (k) ring-roll crusher; (l) hammer mill; (m) dust blower; (n) steam pulverizer; (o) batch ball mill; (p) continuous ball mill; (q) tube mill; (s) rod mill; (t) vibrating ball mill; (u) rubbing mill.

charge is fed under the mullers with plows, and the fines sift out through the screen plates in a continuous process. They are made in several sizes; a large machine might be 10 feet in diameter and grind 50 tons per hour.

Wet pan. This is similar to the previous machine, but is used for batch mixing and grinding of wet mixes (i).

Hammer mill. This mill (l) is used for brittle materials. A series of hammers continually strikes the feed, reducing it until it can pass out through the screen plate. Capacities up to 150 tons per hour can be reached.

Ring roll crusher. This mill as shown at (k) is similar to the preceding one except for the design of the hammers.

Disintegrator (*dust blower*). This machine (m) is a heavy centrifugal fan used for breaking up lump clay or filter cakes in preparation for dust pressing.

Steam pulverizer. This mill (n) is unique in that there are no moving parts. The comminution takes place as a result of the impact and abrasion of the particles on each other and on the lining. It is capable of fine grinding, since there is a classifying action. However, the steam consumption per ton is quite large.

Ball mills. These mills are used extensively in the ceramic industry for fine grinding such materials as quartz, feldspar, and cement clinker. They may be used either wet or dry. The simplest form (o) is a hollow cylinder lined with stone or porcelain and containing hard balls. As the cylinder revolves, the balls tumble over one another and grind the material between them. The grinding efficiency depends on many factors, such as rate of turning, size of the balls, specific gravity of the balls, and amount of charge.

Another type of mill may be used continuously, as shown in (p). The conical shape segregates the balls of different sizes for efficient grinding. Still another type known as a tube mill (q) is used for continuous milling.

The rod mill, (s), employs rods rather than balls. It is widely used for ores, but only to a small extent for ceramics because of the iron contamination. A new type of mill (t) vibrates the mass of balls at high frequency to achieve very rapid fine grinding.

Ball mills regularly grind below 200 or 325 mesh and sizes averaging as low as 5 microns may be obtained. A typical mill for feldspar might have a cylinder $7\frac{1}{2}$ ft in diameter and 10 ft long inside. It would require 85 hp, take $\frac{3}{4}$-in. feed and deliver 1 ton per hour of 90 per cent of the material through a 325-mesh screen.

Rubbing mills. This early type of mill is shown in (u). Heavy stones are dragged over a stone base and abrade the wet mix. Such mills have a low power efficiency and are little used in this country.

Size classification

Screens. Size classification is carried out by means of screens down to 120 mesh, or in some cases to 325 mesh. These screens are usually woven from bronze wire; they form a series shown in Table 7–1. Coarser screens are often made from perforated plates or bars as shown in Fig. 7–4. The taper illustrated here does much to keep the screen from clogging. Recently a considerable increase in the efficiency of any screen

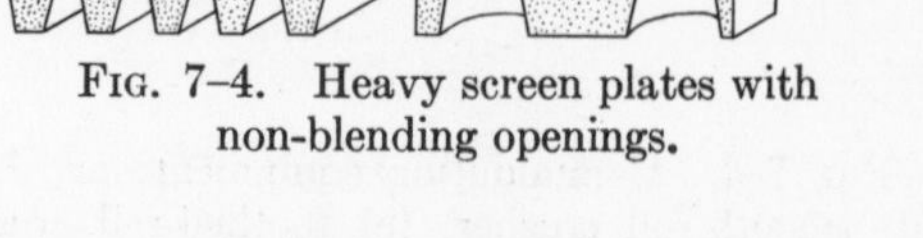

Fig. 7–4. Heavy screen plates with non-blending openings.

Table 7-1
Tyler Screen Series

Openings, inches	Openings, millimeters	Mesh per lineal inch	Diameter of wire, inches
1.050	26.67		0.148
0.883	22.43		0.135
0.742	18.85		0.135
0.624	15.85		0.120
0.525	13.33		0.105
0.441	11.20		0.105
0.371	9.423		0.092
0.312	7.925	2½	0.088
0.263	6.680	3	0.070
0.221	5.613	3½	0.065
0.185	4.699	4	0.065
0.156	3.962	5	0.044
0.131	3.327	6	0.036
0.110	2.794	7	0.0328
0.093	2.362	8	0.0320
0.078	1.981	9	0.0330
0.065	1.651	10	0.0350
0.055	1.397	12	0.0280
0.046	1.168	14	0.0250
0.0390	0.991	16	0.0235
0.0328	0.833	20	0.0172
0.0276	0.701	24	0.0141
0.0232	0.589	28	0.0125
0.0195	0.495	32	0.0118
0.0164	0.417	35	0.0122
0.0138	0.351	42	0.0100
0.0116	0.295	48	0.0092
0.0097	0.246	60	0.0070
0.0082	0.208	65	0.0072
0.0069	0.175	80	0.0056
0.0058	0.147	100	0.0042
0.0049	0.124	115	0.0038
0.0041	0.104	150	0.0026
0.0035	0.088	170	0.0024
0.0029	0.074	200	0.0021

has been made possible by heating the screen wires by electrical conduction. This prevents clogging and increases the life of the screen. Silk screens are often used for fine abrasive materials, since they last longer than metal screens.

In order to provide a constant flow of material through the screen, the mesh must be vibrated in some way. Many types of vibrators are used, some of which are shown in Fig. 7–5.

Screening may be done wet or dry, but for the very fine sizes wet screening is generally more efficient. Several screens are

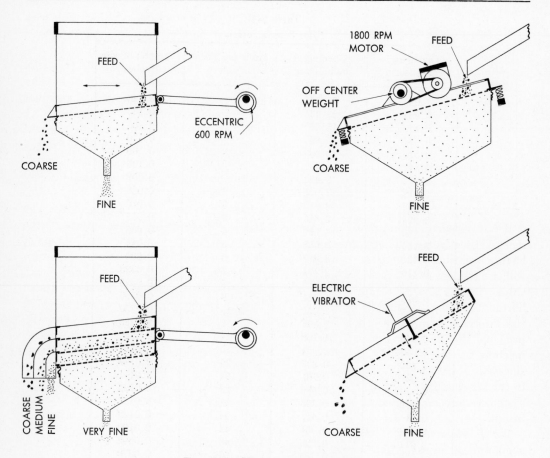

FIG. 7-5. Various types of screen.

often superimposed, with the coarse ones at the top, to separate the feed into several classes and to prevent over-loading the finer screens.

Air classifiers. When the final product is used in the dry condition it is generally more economical to dry grind to avoid the expense of final drying. Therefore, such products as flint and feldspar are dry ground. During the grinding operation, if it is continuous, the fines are swept out of the mill by a current of air or carried away by a mechanical conveyor. They are put through a classifier which takes out the coarse mate-rial for return to the mill; then the fines go into storage.

The rate of settling of fine particles in air or water is given by Stokes' Law as follows:

$$v = \frac{2}{9} \frac{g(\rho_1 - \rho_2)}{\eta} r^2,$$

assuming spheres or equivalent spheres. In this equation, ρ_1 = density of the solid particles and ρ_2 that of the fluid particles, r is the radius of the particle, and η is the viscosity of the fluid. Values of v arrived at by using the appropriate constants are shown in Table 7-2.

Table 7-2

Settling Velocity of Particles in Air and Water

Particle size in equivalent spherical diameter in microns	Settling velocity in cm per second	
	In air	In water
1	0.0077	0.000082
2	0.031	0.00032
5	0.19	0.0020
10	0.77	0.0081
25	4.8	0.050
44 (325 mesh)	15	0.16
74 (200 mesh)	42	0.44
104 (150 mesh)	81	0.87

If a dust-laden column of air or water is rising vertically in a cylinder, those particles having a settling velocity less than the fluid will rise to the top and those with a greater value will stay at the bottom. After equilibrium is reached, a close division into two size fractures is accomplished. This is the principle of the elutriator, an important machine for classification. Often the classification is hastened by centrifugal action in the air classifier. A typical air classifier system connected with a ball mill is shown in Fig. 7–6(a).

Water classification. This method is similar to air classification except that the supporting medium is water. It is used extensively for taking the coarse material from kaolin. In Europe much of the flint and feldspar is wet ground and continuously classified by settling in water.

In England, the china clay is washed by passing a slip of about 2 per cent solid concentration at a velocity of 1 to 2 feet per second through long troughs called "micas" that contain riffles where the grit and mica settle out. The clay is then settled in large tanks and dried out or filter-pressed to remove the water.

Another washing method consists of passing the well-blunged and deflocculated slip through a bowl classifier, as shown in Fig.

7–6(e). This is a large, shallow tank with a conical bottom. As the grit settles, a slowly moving scraper pushes it out the center hole in the bottom, while the clean slip passes over the edge into a launder where it is flocculated and fed to a similar but larger classifier for dewatering. A classifier 25 feet in diameter with a slip of 2 per cent solids will degrit 50 tons of clay a day with about 3 per cent clay loss.

Still another classifier for washing clay is the continuous centrifuge. Fig. 7–6(c) shows a cross section of this machine. The grit is thrown to the inner surface of the bowl and scraped out the small diameter end of the bowl, while the clean slip is thrown out of small holes at the large diameter end. A centrifuge 36 inches in diameter driven by a 15 hp motor at 1400 rpm will degrit 10 tons of dry clay per hour. The clean slip may then be dewatered by passing it through another centrifuge running at higher speeds. Well engineered gearing is needed in this machine to give the slow differential speed to the scraper. The flow sheets at the end of the chapter show some of these processes.

Disintegration

It is necessary to break down clays to their ultimate particles as a preliminary to any washing process, or sometimes to prepare a slip from washed clay filter cakes. This is done in a wet mixer called a blunger. The old types were large cylindrical tanks in which slowly moving paddles were mounted on a vertical shaft. The more modern types use a high speed motor that circulates the slip and rapidly breaks down lumps, as shown in Fig. 7–6(d).

Chemical treatments

Ceramic raw materials, except for chemicals used in glazes or special refractories, are seldom chemically treated because of

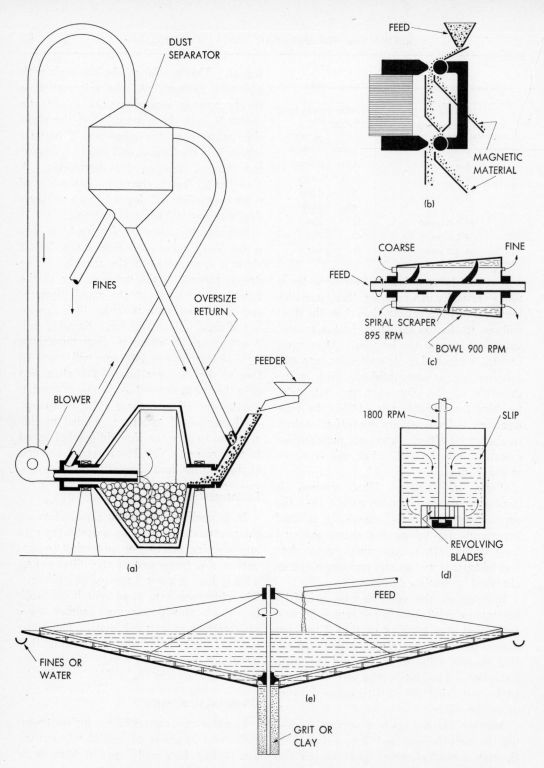

FIG. 7–6. Classification operations: (a) continuous ball mill with air separator; (b) magnetic separator; (c) dewatering centrifuge; (d) high speed blunger; (e) Dorr type bowl classifier.

66

the added cost. An exception is sea water magnesite, which is now produced in large quantities as shown in the flow sheet of Fig. 7–12.

In Europe, glass sands are sometimes treated to remove iron, while here some of the paper clays are bleached with zinc hyposulphite.

The materials for some of the special refractories are produced chemically, for example, alumina, zirconia, and beryllia, but since these processes serve the ceramic industry to only a small extent they will not be described here.

Magnetic separation

This method is widely used, especially to remove iron or iron minerals from feldspar. A cross section of a high-intensity magnetic separator is shown in Fig. 7–6(b). Minerals with rather low susceptibilities may be taken out in the intense field generated. A relatively small unit will handle 2 to 4 tons of feldspar per hour with a power consumption of only 1 kw-hr per ton. It is, of course, necessary to grind the mineral fine enough to unlock the magnetic grains. On the other hand, very finely ground material does not pass through a magnetic separator easily because of caking and uneven flow.

Froth flotation

In the last twenty-five years the methods of flotation have revolutionized the treatment of ore minerals. This method consists of mixing the finely ground, water suspended ore with a frothing agent, whereby there is a differential adsorption by the bubbles on the ore particles and the gangue particles. One or the other floats to the surface and is removed as shown in Fig. 7–7.

In the ceramic field, flotation has not been used to a great extent in actual production,

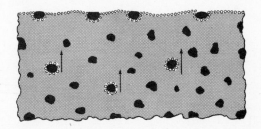

Fig. 7–7. Froth flotation due to selective adsorption of bubbles on one of the minerals in suspension.

although considerable research has been conducted with various minerals. At present the feldspar industry uses flotation to some extent in removing quartz from feldspar. Also, mica flakes are floated from sand and quartz as a kaolin by-product. It seems reasonable to expect that during the next ten years flotation methods will be used more and more in the industry.

Filtering

Water removal by filtering is a common practice in the ceramic industry, since it tends to take out soluble salts. The plate-type filter press is generally used, but continuous filters of the drum type are desirable for large production.

Drying

Bulk drying is still carried out in open sheds under natural conditions, but in modern plants it is done in rotary dryers, generally of the indirect type shown in Fig. 7–8, to prevent contamination by combustion products.

Some lump material, such as filter cakes, may be dried on cars in humidity controlled continuous driers or on conveyor-type driers.

Spray drying of clay slip, done by injecting it into a hot chamber, has been used for some clays, but a very careful temperature control is needed to prevent breaking down

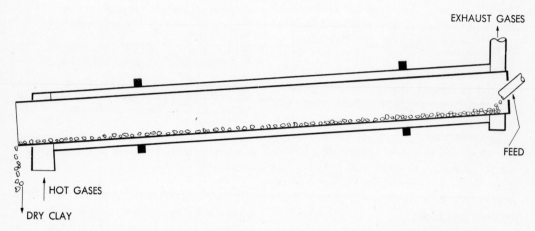

EXHAUST GASES

FEED

HOT GASES

DRY CLAY

FIG. 7–8. A cross section of an indirectly heated rotary dryer for handling lump clay.

the clay structure and thus reducing subsequent plasticity.

Storage and handling

Ceramic industries, such as a whiteware pottery, that use relatively small amounts of raw materials can afford to store a supply sufficient for weeks or even months. In these cases material is shipped from some distance and an interruption of flow must be guarded against. On the other hand, refractory and heavy clay products manufacturers usually mine their clay close at hand, and use such large quantities that a day's storage may be all that is economical.

In the European potteries storage serves another purpose, that of aging. In many plants the clays are left outdoors for a year or more to freeze and thaw or to dry and be wetted by rains, processes which undoubtedly improve their working properties.

The trend in handling these raw materials is toward substituting conveyors for hand labor as far as possible. There is not space here to go into the many types available, but a visit to a modern plant will show that many labor-saving devices are used.

Flow sheets. A number of flow sheets for material processing are shown in Figs. 7–9 to 7–14.

REFERENCES

BOWLES, O., *Asbestos — Milling, Marketing and Fabrication.* U. S. Bur. of Mines, Inf. Circ. 6869, 1935.

BURGESS, B. C., *Milling Feldspars.* U. S. Bur. of Mines, Inf. Circ. 6488, 1931.

GARVE, T. W., *Factory Design and Equipment and Manufacture of Clay Wares,* 2nd ed. T. H. Randall and Co., Wellsville, N. Y., 1945.

KOENIG, E. W., "Froth Flotation as Applied to Feldspar." *Ceramic Age* 47, 192, 1946.

LADOO, R. B., *Feldspar Mining and Milling.* U. S. Bur. of Mines, Repts. Invest. 2396, 1922.

RIDDLE, F. H.. "Mining and Treatment of the Sillimanite Groups of Minerals and Their Use in Ceramic Products." *Trans. A. I. M. E.* 102, 131, 1932.

WEIS, J. H., "Granular and Ground Feldspar with a Uniformly Low Iron Content." *J. Am. Ceram. Soc.* 14, 413, 1931.

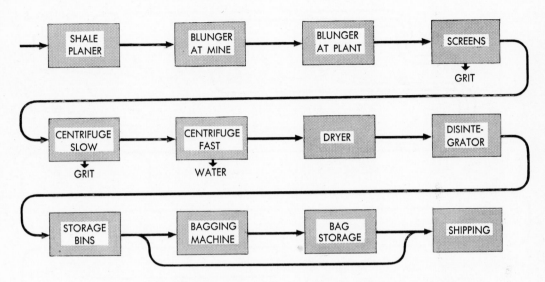

FIG. 7–9. Flow sheet illustrating the preparation of washed Georgia kaolin.

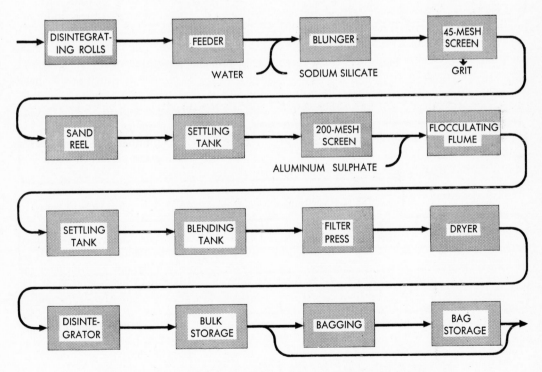

FIG. 7–10. Flow sheet of European method of producing washed kaolin.

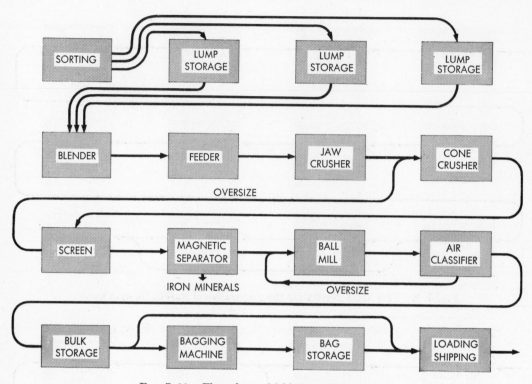

Fig. 7–11.   Flow sheet of feldspar milling process.

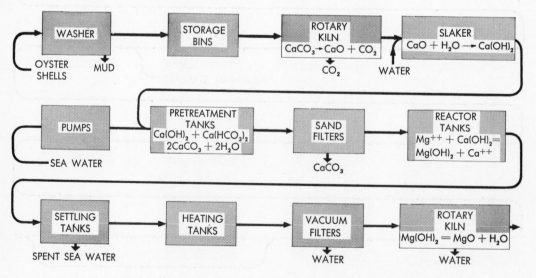

Fig. 7–12.   Flow sheet showing the lime process of making magnesia from sea water.

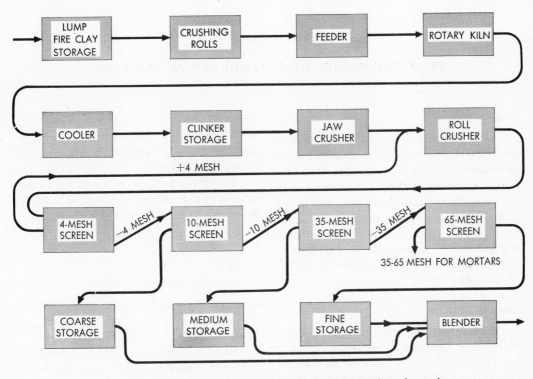

FIG. 7–13. Flow sheet for the production of sized grog for refractories.

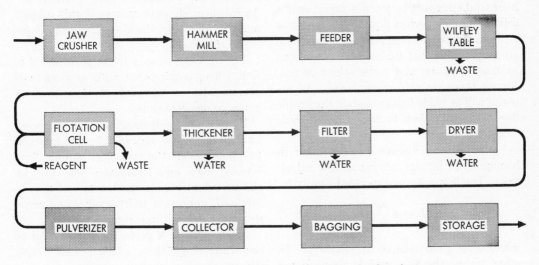

FIG. 7–14. Flow sheet for the production of talc by the flotation process.

CHAPTER 8

FLOW PROPERTIES OF CERAMIC PASTES AND SLIPS

Introduction

A vital characteristic of clays and of most ceramic bodies is the nebulous quality known as plasticity. This is the quality that permits the forming of the piece, often in very intricate shapes, and then keeps the forces of gravity or the shocks inherent in manufacture from deforming it. This chapter will attempt to discuss the laws of flow in ceramic materials and to advance some explanation for plastic behavior.

Laws of flow

There are several types of flow that are well understood, such as viscous and elastic flow; in addition, there are a myriad of substances that do not see fit to follow these simple laws, but exhibit a flow of considerable complexity. Indeed, it is sometimes hard to distinguish between a liquid and a solid. While a perfect liquid is defined as a medium incapable of supporting shear, there are suspensions that under some conditions have a yield point and under others do not. Among these are clay suspensions. An attempt will be made first to explain the simple flow types and then to take up the more complicated ones.

Viscous state. Viscous flow is found in homogeneous liquids moving at low velocities. The flow may be expressed simply by

$$F = k \frac{dv}{dr}, \qquad (8\text{-}1)$$

where k is proportional to viscosity. In other words, the flow rate is proportional directly to the force applied and inversely to the viscosity. Consider the cube, one centimeter on a side, shown in Fig. 8–1. If a force of one dyne tends to shear the cube at a velocity of 1 cm per sec, then the viscosity of a liquid making up the cube is unity. The dimensions of viscosity are given by

$$\eta = \frac{F}{\dfrac{dv}{dr}} = \frac{\text{dyne} \cdot \text{cm}^{-2} \cdot \text{cm}}{\text{cm} \cdot \text{sec}^{-1}}$$

$$= \text{dynes} \cdot \text{cm}^{-2\cdot} \cdot \text{sec}, \qquad (8\text{-}2)$$

or in dimensions of mass, length and time,

$$MLT^{-2}L^{-2}T = ML^{-1}T^{-1}. \qquad (8\text{-}3)$$

When considering any physical property measured quantitatively, it is always well to determine the units of that quantity by dimensional analysis so that both sides of an equation will be dimensionally compatable. An excellent short treatment of this subject may be found in Mark's *Mechanical Engineers' Handbook.*

The unit of viscosity is the *poise* and it is well to keep in mind that water at room temperature has a viscosity of close to 0.01 poise, or one centipoise.

If more and more shearing force is applied to a fluid, the velocity of flow will increase to a point where the viscous or laminar flow breaks down and turbulence sets in. In the turbulent region another law of flow is followed, important in hydraulics and aerodynamics but not in ceramics.

Elastic state. A perfectly elastic solid is one which obeys Hook's Law — that is, the deformation is proportional to the deforming

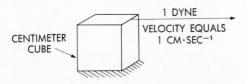

FIG. 8–1. Cube of fluid undergoing shear.

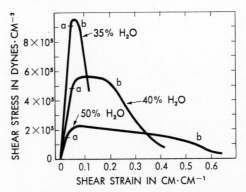

FIG. 8–2. Stress-strain diagram for a plastic clay.

force up to the breaking point. When the deforming force is removed, the solid resumes its original form. There are, perhaps, no perfectly elastic materials in nature, but many solids approach this condition so closely that they may be considered such for all intents and purposes.

Plastic state. Unfortunately, many materials, including most ceramic materials, do not have the simple flow properties described for fluids and elastic solids, but instead approach a complexity that makes the subject very difficult to analyze. The study of the flow in such materials as rubber, fats, waxes, paints, ceramic pastes, clay slips, and even metals is called rheology.

The various types of flow are summarized clearly by Scott Blair. The types that interest us in ceramics are the visco-elastic (glass) and plasto-elastic (clay pastes). The former will be considered in Chapter 13, while the latter will be treated here in some detail.

When a piece of clay paste is stressed with an increasing force, a stress-strain diagram like that in Fig. 8–2 will be obtained. Up to the yield point a, the flow is elastic. Should the stress be released after a very short time interval, the original size would be regained. However, should the stress be maintained for a long time, some of this ability to return would be lost, probably by migration of water from the highly stressed parts to those with lower stress. Carrying the stress beyond the yield point produces plastic flow and allows a considerable deformation of the piece before cracking appears at b. There are, then, two features of importance in this diagram; first, the yield value and second, the extension at breaking.

A workable clay paste should then have a yield value high enough to prevent accidental deforming and have an extension large enough to allow forming without fracture. For a given clay paste these two features are not independent of each other, for by varying the water content it is possible to increase either one. At the same time, the other decreases, as is shown in the stress-strain diagrams of Fig. 8–2. We may then say that the workability may be approximately evaluated by the product of yield point and maximum extension, and that a given water content produces a maximum in this value as shown in Fig. 8–3.

In general all plastico-elastic materials consist of at least two phases, solid and fluid; for example, clay pastes, paints, plasticene, and even metals have hard crystals and soft inter-grain material.

Flow of dispersed suspensions. The flow of clay-water suspensions is of particular interest to ceramists, so some space will be devoted to its consideration. When such a suspension is sheared at a uniform rate and the shearing force measured, a value of con-

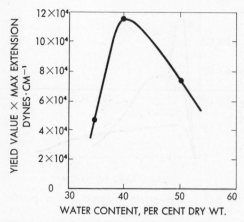

FIG. 8–3. Workability of a clay paste at different water contents.

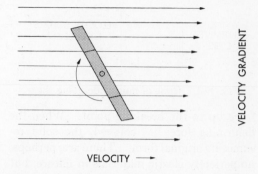

FIG. 8–4. A plate-like particle revolving in a fluid undergoing shear.

sistency which we may call apparent viscosity results from equation (8–2). This is a convenient term to use.

It has been found that the important variable in the dispersed clay-water suspension is the concentration of solids. The other variables such as the sizes and shapes of the particles have less effect.

As the suspension is sheared the disklike clay particles revolve about an axis in the plane of shear, as shown in Fig. 8–4. This rotation absorbs energy in itself, but should the particles approach each other due to a more concentrated suspension, the mutual interference will cause further energy absorption.

It may then be assumed that the total apparent viscosity of the suspension will be due to three contributing factors: first, the energy absorbed in the water itself; second, the energy absorbed by the individual particle; and third, the energy of the particle collisions. Then we may write

$$\eta_s = \eta_l(1 - C) + kC^n + k_1C_s^m, \quad (8\text{–}4)$$

in which n is found by experiment to be unity, m is 3, and the values of k and k_1 are

0.08 and 7.5 respectively for dispersed suspensions. To visualize this suspension, a section of it has been drawn to scale in Fig. 8–5 for a volume concentration C of 0.05 and 0.28. It will be seen that in the dilute suspension each particle has room to turn with little chance of interfering with its neighbor. When C increases to a point where the center to center distance reaches the diameter of the disk, as in the more concentrated suspension, then the viscosity increases with great rapidity. In other words, the first term of equation (8–4) is the important one for very dilute suspensions, the second term for medium concentrations, and the third term for high concentrations.

In the previous discussion the shear rate was considered a constant, but should this value be varied with everything else constant, it would be expected that η_s would be invariable as in the case of Newtonian fluids. This is not the case, however, for suspensions of nonisotropic particles such as clay. There is some evidence that suspensions of spherical particles behave like fluids, but the story is by no means complete.

In Fig. 8–6 are shown curves of η_s for various rates of reciprocal shear, $1/(dv/dr)$. It will be seen that the apparent viscosity decreases with an increasing shearing rate. If

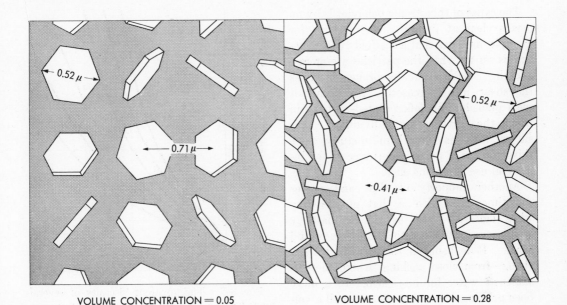

VOLUME CONCENTRATION = 0.05   VOLUME CONCENTRATION = 0.28

FIG. 8–5.   Thin layer of a suspension of monodisperse kaolinite particles at two concentrations.

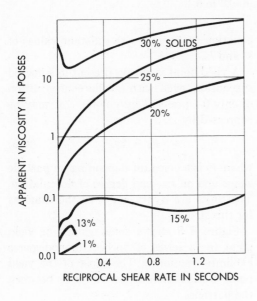

FIG. 8–6.   Influence of shear rate on apparent viscosity.

the curve is extended to zero in the plot (infinite shear rate), a value is obtained called the basic apparent viscosity, which is a convenient value to use for all suspensions.

The property of decreasing apparent viscosity by increasing the shear rate is called thixotropy.   A quantitative measure suggested by Goodeve are the slopes of the curves of Fig. 8–6 at the origin of the plot, or at infinite shear.   The coefficient of thixotropy of suspensions may be expressed by

$$\theta = k_3 C + k_4 C^3, \qquad (8\text{–}5)$$

where k_3 and k_4 are constants depending on the particle size and shape and on the degree of flocculation.   Some authors consider thixotropy to be a change of apparent viscosity with time at constant shear rate.   It is interesting to note that the unit of reciprocal shear rate is seconds.   The units of θ are dynes $\cdot$ cm^{-2} or $ML^{-1}T^{-2}$.

The cause of thixotropic behavior is by no means clear at present. One theory, for example, assumes that a scaffold-type structure is set up when the particles are at rest and that this is gradually broken down as the shear rate increases. This does not seem plausible for deflocculated systems in which the particles are individuals. This subject must have much more study, since thixotropy is of great importance in the industrial use of casting slips.

Another property of slips and pastes sometimes met with is that of stiffening when agitated. This is called dilatancy and is encountered with coarse clays and sands. It is believed to be caused by a change from close packing, where there is sufficient water for lubrication, to a more open packing where there is too small a volume of water to fill the voids.

Rheopecsy is a term used to describe the gelation produced in a colloid by gentle agitation.

Another property of suspensions is the yield point, or the maximum shearing strain that may be supported before continuous shear starts. It has not been possible with our most sensitive instruments to detect any yield point in deflocculated suspensions.

Flow of flocced suspensions. A flocced suspension has a much higher coefficient of apparent viscosity than the same suspension when deflocculated. In the former there are groups of particles rather than single ones, and these groups offer great resistance to shear. However, the apparent coefficient of viscosity of the flocced suspension may be expressed by equation (8–4) with different values for k and k_1. Figure 8–7 gives curves of η_s for a series of monodisperse fractions, to show the effect of particle size. In the same way the thixotropy for the flocced suspensions may be expressed by

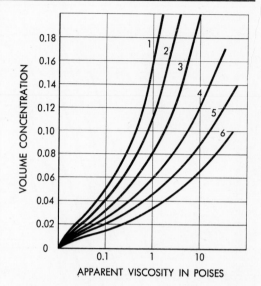

FIG. 8–7. The apparent viscosity of several monodisperse fractions in aqueous suspension (flocced). Figures refer to equivalent spherical diameter in microns: (1) 6.4 to 12.8; (2) 3.2 to 6.4; (3) 1.6 to 3.2; (4) 0.8 to 1.6; (5) 0.4 to 0.8; (6) 0.2 to 0.4.

equation (8–5), but with different values of k_2 and k_3.

A yield point is always present in a flocced suspension, even with a solid concentration of only 0.1 per cent by volume. It may be expressed by

$$F_0 = k_4 C^3, \qquad (8\text{--}6)$$

where k_4 is a constant depending on particle shape and on size and degree of flocculation. The units are dynes·cm^{-2}, or the same as for thixotropy.

Figure 8–8 shows some values of yield point for a series of flocced, monodisperse kaolinite fractions. The cause of the yield point is the action of elastic forces between the particles.

Influence of particle shape on flow of suspensions. A study made with similar frac-

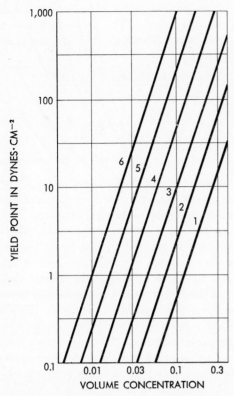

FIG. 8–8. Yield point of flocced suspensions. (The particle size is shown in Fig. 8–7.)

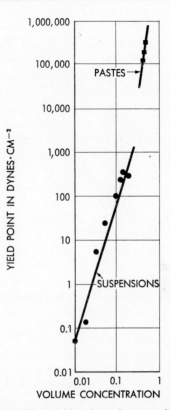

FIG. 8–9. The yield point of suspensions and pastes of No. 2 fraction (See Fig. 8–7).

tions of various particles of increasing elongation indicates that the more isometric the particle in deflocculated suspensions, the lower is the coefficient of apparent viscosity. In other words, it takes more energy to rotate elongated particles than spherical ones. However, when the system is flocced, the particle shape has no influence, as clumps rather than individual particles are encountered.

Mechanism of plasticity

Introduction. The subject of plasticity has been discussed from many points of view, but there is not space here to review the many theories that have been devised. The reader is again referred to Scott Blair for a summary. In this section there will be set forth the facts as we know them, and then the theory will be described that seems the most plausible to the author. However, much more study of this problem is needed before the picture can be made clear.

Difference between suspensions and plastic masses. If the yield point of a monodispersed, flocced suspension is plotted over the widest range of concentration possible for measurement, then the same is done for plastic masses of the same particles and the values placed on a single plot, the curves in

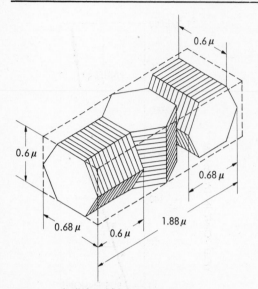

FIG. 8–10. Packing of kaolinite plates in a plastic mass.

8–10 to give 14 water films for each 1.88 microns of length. The linear shrinkage is 5.0 per cent. Therefore, the film thickness is given by

$$\frac{\% \text{ linear shr. of mass} \times 1 \text{ cm}/100}{\text{No. of films per cm}}$$

$$= \frac{.050}{\dfrac{14}{1.88} \times 10^4} = 6.6 \times 10^{-7} \text{ cm}$$

$$= 6.6 \times 10^{-3} \text{ microns.}$$

Of course the film thickness will change with the pressure used in forming the mass, as will be discussed in the next chapter.

Forces acting between the particles. In a clay mass at equilibrium the forces holding the clay particles together must be just balanced by those holding them apart. It is not difficult to measure the repulsion forces for any given separation distance. This is done by pressing the plastic clay between permeable pistons. The clay mass, after pressing, is dried and the water film thickness determined as described previously. With the size of each kaolinite particle known, the force between particles is readily calculated, as will be done in Chapter 10.

It is evident that the force holding the particles together is just equal to the repulsion force. For example, when a piece of plastic clay is pressed between permeable pistons at high pressure, water is squeezed from the mass and the particles come together to the point where their repulsion balances the external pressure. Now if the pressure is released under conditions where no water is available to flow back into the mass, then the films remain at their high pressure thickness. There must be available for holding the particles together some force of the same magnitude as the released pressure.

The only force that would seem to be available is the capillary force at the surface

Fig. 8–9 are obtained. In the first place there is a gap in the sticky range of clay consistency in which there is at present no way of measuring yield point; this is a range which cannot be used in ceramic production. Second, it will be observed that the slopes for the two consistency regions are different. For the slips it is 3 and for the solids 6, which indicates that a different law governs the consistency of the solid; otherwise the points for the solid would follow an extension of the line for the suspensions.

Water film in the plastic mass. If a plastic mass is made up with a monodisperse fraction of kaolinite when the size and shape of the particles are known, it is possible to calculate with considerable exactness the thickness of the water film between the particles. This is done by drying the mass and measuring the linear shrinkage. Taking a particular case, the particles are hexagonal plates, 0.6 micron across the flats and 0.05 micron thick. Statistically, these plates may be considered as set up in Fig.

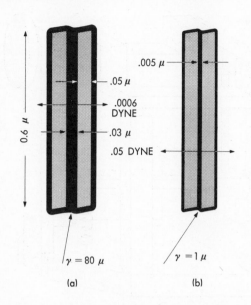

FIG. 8-11. The forces between a pair of kaolinite plates.

lary surface enables the attractive force to just balance the repulsion. The water in the films must be in hydrostatic tension. That this is actually so is indicated by the experiments of Westman, who showed values as high as 880 lb/sq in. for plastic ball clays.

The same reasoning applied to the two kaolinite plates may also be applied to a clay mass with millions of particles. The surface layer of water acts as a stretched membrane, forcing the particles together. As the clay dries out, the water layers between the particles decrease and the surface membrane becomes thinner and pulls down between the particles to exert greater force.

Three consistencies of a clay-water mass are shown in Fig. 8-12, which is a cross section at the surface. A very simple experiment will illustrate the stretched membrane. A toy rubber balloon is filled with dry, pulverized clay, in which case the clay feels like a dry powder. If the balloon is slowly evacuated so that the atmosphere presses on the rubber to hold the clay particles together, a remarkable change occurs; the clay in the balloon now feels just like a plastic clay-water paste.

To sum up, a plastic mass has a yield point that must be exceeded before plastic

of the clay mass. It will be easier to understand this force if we consider the very simple case of two kaolinite plates surrounded by a water film. These plates, drawn to scale, are shown in Fig. 8-11(a) for a thick water film and in Fig. 8-11(b) for a thin film. The radius of curvature of the capil-

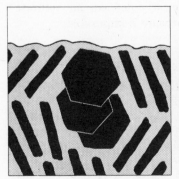
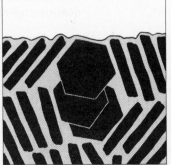

FIG. 8-12. A cross section of a clay paste at three water contents.

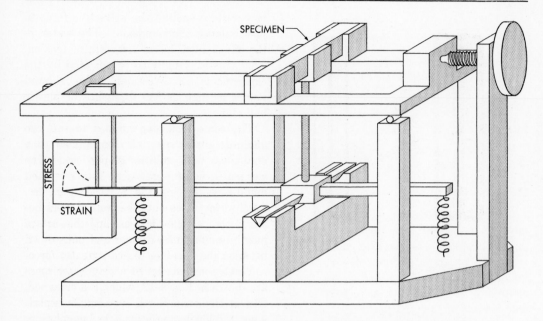

FIG. 8–13. Apparatus for measuring the flow properties of clay pastes in shear.

flow may be initiated. When flow starts it may be carried to some extent before fracture occurs. The more plastic masses have both a high yield point and a long extension before fracture. The yield point is believed to be due to the holding together of the mass by the outer surface film and to the extensibility caused by platelike particles that readily slide over each other, perhaps lubricated by the water films.

Measurement of plasticity

Numerous instruments have been devised for the measurement of plasticity. There is not space here to describe them, but a simple device is shown in Fig. 8–13 for obtaining a stress-strain diagram for pastes in shear. In the case of suspensions, the well-known MacMichael type revolving cup viscosimeter is most useful. Until plasticity can be more clearly defined, an accurate quantitative measure is impossible.

REFERENCES

Bingham, E. C., *Fluidity and Plasticity.* McGraw-Hill Book Co., Inc., New York, 1922.

Goodeve, C. F., "A General Theory of Thixotropy and Viscosity." *Trans. Faraday Soc.* **35,** 342, 1939.

Green, Henry, *Industrial Rheology and Rheological Structures.* John Wiley and Sons, Inc., New York, 1949.

Nadai, A., *Plasticity.* McGraw-Hill Book Co., Inc., New York, 1931.

Norton, F. H., "Fundamental Study of Clay: VIII, A New Theory for the Plasticity of Clay-Water Masses." *J. Am. Ceram. Soc.* **31,** 236, 1948.

Scott Blair, G. W., *A Survey of General and Applied Rheology*, 2nd ed. Pitman and Sons, Ltd., London, 1949.

CHAPTER 9

PLASTIC MASSES

Introduction

In this chapter the properties of various plastic masses will be discussed.

Clay-air-water system

Low molding pressure. This system is a most fundamental one and concerns all plastic masses. The diagram in Fig. 9–1 gives a very simple picture of the relative volumes of the three phases as the water content is increased. In this diagram the true volume of the clay particles is taken as unity and the bulk volume of the clay as 1.6. Water is added to the system by mixing it thoroughly with the clay, and the mass is then pressed lightly into a definite shape.

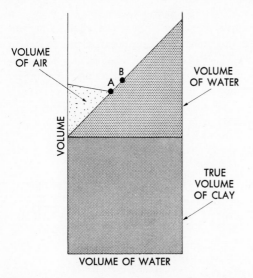

FIG. 9–1. Clay-water-air system at low molding pressures.

The line bounding the water volume in the diagram is, of course, at a slope of 45°. The volume of air in the system added to the water volume gives the total, or bulk volume curve.

As water is added to the system it gradually displaces the air with little change in bulk volume, although a slight decrease is usually noted as the lubricating action of the water permits a little better packing of the particles. When point A is reached, all the air is completely replaced by water. This water is in the pores, but has not built up water layers between the particles, for at this point there is no appreciable drying shrinkage. As more water is added, the total volume increases until the saturation point B is reached, beyond which any additional water is squeezed out at this particular molding pressure. The region from A to B is the plastic range at this molding pressure. At A the body has a high yield point and low extension, but as B is approached, the yield point decreases while the extension and the drying shrinkage increase. It should be kept in mind that a plastic mass is a two-component system — clay and water only.

Before going into this subject further, it is advisable to stop and consider a very strange fact; namely, that the water gradually added to the clay first fills the pores and then, only when the pores are full, builds up the interparticle films. In other words, at a point just to the right of A on the diagram there exists a condition like that in Fig. 9–2(b), nòt what might be ex-

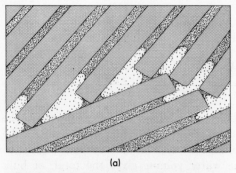

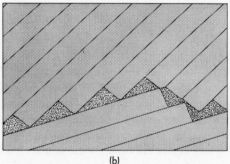

FIG. 9–2. Schematic cross section of clay mass.

pected in Fig. 9–2(a). The latter condition would occur if the clay surface had a strong attraction for water, as is often believed. How can this unexpected behavior be explained? There seems to be no answer yet to this baffling question.

High molding pressure. The clay-air-water system shown in Fig. 9–1 may be repeated at a higher molding pressure, such as that used in the dry pressing process. The diagram for these conditions is shown in Fig. 9–3. The field boundaries for the clay and water are, of course, unaltered, but the area representing the air is reduced because of the closer packing of the particles under high pressure. Points *A* and *B* both come at lower water contents.

Masses containing minerals other than clay

While wet clay is usually thought of as *the* plastic material, other minerals, when finely ground, exhibit this property to some extent. While it is generally believed that platelike minerals tend to show greater plasticity than those with chunky particles, there is no definite evidence to show that any mineral, if ground so that its average dimension is in the range of thickness of the kaolinite plates, 0.1 to 0.01 micron, will not have claylike plasticity. Even quartz, with no definite cleavage, develops some plasticity when finely ground.

Particle packing. As clay alone often has a large drying shrinkage and undesirable firing properties, it is common practice in many branches of the ceramic industry to add a nonplastic like silica, feldspar, or grog (hard fired clay). The particle size of the nonplastic has an important influence on the properties of the body, so some space will be devoted here to the influence of particle size distribution.

If the nonplastic is of one grain size, the volume of pores for crushed grains will be close to 45 per cent of the bulk volume. In

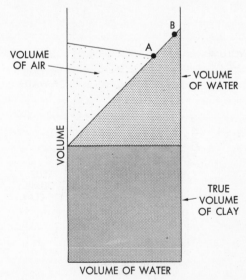

FIG. 9–3. Clay-water-air system at high molding pressure.

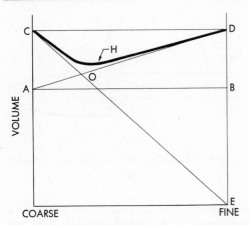

FIG. 9–4. Packing diagram of coarse and fine grog.

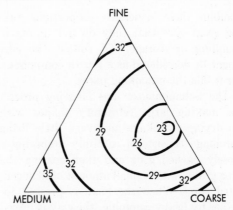

FIG. 9–5. Packing diagram of a three-component grog system. Contour figures represent porosity based on bulk volume.

some cases it is desirable to have a lower pore volume in order to get the maximum possible amount of nonplastic in a given volume. By using two sizes of grog mixed together, the pore volume may be reduced as shown in Fig. 9–4. Line AB represents the true volume of 100 gm of nonplastic that consists of various mixtures of fine and coarse. D will be the bulk volume of the fine fraction and C of the coarse fraction, and line CD represents the bulk volume of the unmixed fine and coarse components. However, if the two are thoroughly mixed, the bulk volume will shrink to line COD, since the fine particles will fit into the pores of the larger ones. The minimum volume comes at O, where the pores in the larger fraction are just filled by the smaller one. The student should thoroughly understand this diagram and know, for example, the reason that line CO is prolonged to the base line while line DO is prolonged to the true volume line. It is helpful to take some 250-cc graduates, fill them with mixtures of two sizes of crushed grog, and then vibrate them until the bulk volume becomes a minimum. The packing of the particles can readily be seen during this operation.

The diagram of Fig. 9–4 is made up for two fractions with an infinite ratio between the fine and the coarse. Actually a ratio of ten is commonly attained, which gives a bulk volume curve shown by curve H.

If three components are used, still closer packing may be attained; the fine fraction goes into the pores of the medium fraction, which in turn fills the pores of the coarse fraction. Theoretically, still denser packing could be obtained by four or five components; actually, this does not follow, since it is impossible to keep a large size ratio between components if more than three are used.

Figure 9–5 shows the packing density for three sizes of crushed grog mixed in all proportions. The densest packing comes at 50 per cent C, 10 per cent M, and 40 per cent F, with a porosity of 22 per cent based on the bulk volume, not a great gain over the 25 per cent figure for the fine and coarse mixture alone.

Practical application of particle packing. So important is this question of particle sizing in the refractories industry that many modern plants separate the crushed grog into size fractions by screening and then combine these in definite proportions, taking great care that they do not unmix in handling or storage. Of course, the clay might be considered as a fourth component, for it fills the remaining pores.

The Schaunhauser and Giessing process for making large refractory shapes with no drying shrinkage and very little firing shrinkage consists of carefully preparing a closely packed grog and then coating the grog grains with a small amount of deflocculated slip. When the mass is pressed together by heavy ramming, the clay is forced into the few voids left. The grog particles make such close contact that little or no subsequent shrinkage can result.

The whiteware body is, perhaps by accident, a closely packed three-component system with the flint and feldspar as the coarse, the kaolin as the medium, and the ball clay as the fine component. It has been shown that molded densities even higher than those found in practice may be attained by a more careful selection of the components, thus reducing shrinkage and warping in the kiln.

Effect of vacuum treatment on plastic masses

It was stated earlier in this chapter that there was no air in the usual plastic mass. However, it is difficult to remove the last traces of adsorbed air by ordinary means. This trace of air is likely to cause inhomogeneities in the structure and a reduction in density. Therefore, the vacuum treatment of plastic bodies has become quite common, and the following section will be devoted to the methods used and the result obtained.

Means for vacuum treatment. The air in plastic masses diffuses through the structure with extreme slowness. Hence its rapid removal entails breaking the clay down into

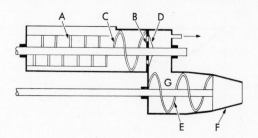

Fig. 9–6.   Simplified cross section of a vacuum auger.

small pieces and placing these pieces in a vacuum while they are being deformed and consolidated.

This operation is carried out in a vacuum auger, shown diagrammatically in Fig. 9–6. The clay is fed into chamber A with the correct amount of water; it is thoroughly worked and then forced through the perforated plate B by the auger C, very much in the manner of a meat grinder. As the small pencils of clay emerge from the plate they are still further broken down by the blade D. They then fall into chamber G, which is maintained at a vacuum of about 26 inches of mercury. In this chamber the clay mass is worked and consolidated into a homogeneous air-free structure and extruded by the auger E through the die F.

Vacuum-treated clay is in general more plastic than untreated clay, and this treatment will permit the use of very short clays which might not otherwise be workable. A considerable proportion of the plastic clay bodies made in this country are now vacuum treated. It should not be concluded that this is the only way to remove air; however, it is the most efficient. Careful hand wedging, together with aging periods, produces a very homogeneous body, as do the roller kneading machines used in European potteries.

Effect of aging on the plastic mass

It has been the practice in the older potteries and clay works to let the clay and water age for considerable periods of time with perhaps intermediate periods of reworking. It is said that the ancient Chinese potter made up enough porcelain body for his son to use during his lifetime. Even now many European factories age their bodies for weeks or even months. The aging process is apparently of a twofold nature; one is the complete wetting of the clay and the other a bacterial action of which little is known. In this country we cannot afford the time or inventory tie-up of long aging in our production processes. The vacuum treatment somewhat takes its place.

Effect of adsorbed ions

As the nature and effect of adsorbed ions will be discussed in the next chapter, only one point will be brought out here; that is, adsorbed ions that tend to deflocculate lower the yield point so much that a good workable mass is impossible. Most plastic clays contain adsorbed calcium ions, a condition that produces a high yield point.

REFERENCES

MACY, H. H., "Clay-Water Relationships." *Proc. Phys. Soc.* (London) **52,** 625, 1940.

WESTMAN, A. E. R., "The Packing of Particles." *J. Am. Ceram. Soc.* **13,** 767, 1930.

WILLIAMSON, W. O., "The Physical Relationship between Clay and Water." *Trans. Brit. Ceram. Soc.,* L, 10, 1951.

CHAPTER 10

CASTING SLIPS

Introduction

The casting slip is a suspension of ceramic materials in water, just thin enough to pour. It is poured into dry plaster molds which draw enough water out from the slip to raise its yield point above that needed to support the cast against forces of gravity. It is customary to deflocculate the slip to reduce the water content and to give firmer casts. Casting is an ancient art, used as early as the year 1700 in Europe. However, the first application of deflocculants is mentioned by Brongniart in 1844.

Elements of colloidal chemistry

Nature of colloids. A colloid is any type of material so finely divided that its surface effects become important when compared with the bulk effects. The upper limit for colloidal materials may be on the order of 5 microns, and the lower limit approaches molecular size. Colloidal particles may be lyophobic (shunning water) or lyocratic (attracting water). Clays and nearly all other ceramic materials are in the latter class.

Lyosphere. Each colloidal particle suspended in water is believed to be surrounded by a lyosphere, an envelope of water molecules attracted to the particle surface. As explained before, the clay particle may have adsorbed ions on the surface held either by broken bond energy or by the surface energy of the complete lattice. These adsorbed ions are believed also to be hydrated by attracting water molecules to themselves, but

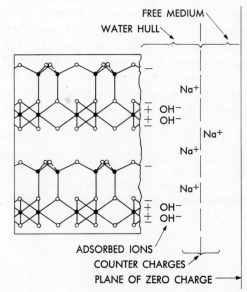

Fig. 10–1. Broken edge of a kaolinite crystal showing the lyosphere.

there is no exact data on ion hydration. One of the most generally accepted theories is that of Gouy-Freundlich. It postulates a lyosphere in two layers, caused by selective adsorption, as shown diagrammatically in Fig. 10–1. The net charge on clay particles as far as can now be determined is always negative, and therefore they will always travel to the anode (positive); by measuring the rate of travel, the zeta potential or charge across the lyosphere may be calculated. From this and the dimensions of the particle, the total charge may be approximately determined. This value for 0.3 micron diameter particles is shown in Fig. 10–2. In the hydrogen or flocculated

86

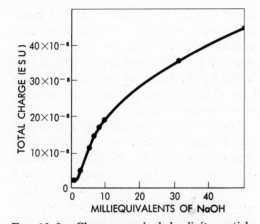

Fig. 10–2. Charge on a single kaolinite particle.

state the charge is substantially zero, but with increased NaOH the charge increases rapidly.

Deflocculation of slips

As in all precise work with clay, it is advisable to carry out the first experimentation on clean, monodisperse fractions of kaolinite. Figure 10–3 shows the curve of apparent viscosity of a clean Florida kaolin suspension of 16 per cent solids as the NaOH content is increased. At the origin the clay is free of all adsorbed ions except H^+ and $(OH)^-$, but as NaOH increases these hydrogen ions are gradually replaced by Na^+. There is no marked change in viscosity until the base exchange capacity is reached at 3.7 milliequivalents, where all the adsorbed hydrogen ions are replaced. As soon as there is any excess NaOH in the suspension, the viscosity drops suddenly to only 1/200 of its original value, a remarkable change.

On the same plot is shown a curve of hydrogen ion concentration or pH. This value shows a sudden rise to the basic condition at exactly the same content of NaOH that is required to drop the viscosity. In other words, the suspension must have free Na^+ and OH^- to bring about deflocculation.

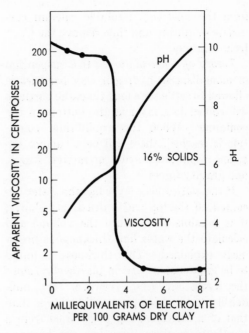

Fig. 10–3. Viscosity and pH of a clean, monodisperse fraction of kaolinite.

A fairly complete survey of deflocculants has shown that they must have two characteristics, one a basic reaction, and the other a monovalent cation. Therefore, all deflocculants are salts of the alkali metals or ammonia, such as sodium carbonate, sodium silicate, and sodium hydroxide, that hydrolyze to give a basic reaction.

In Fig. 10–3 it was shown that sodium hydroxide was an excellent deflocculant for a clean clay. However, practice has shown that it is not satisfactory for use with commercial clay slips, whereas sodium carbonate and sodium silicate are excellent. This may be explained by the fact that commercial clays have adsorbed calcium ions which form the slightly soluble calcium hydroxide with sodium hydroxide. Thus divalent ions are produced in solution, a fact that hinders deflocculation. On the other hand, both sodium carbonate and sodium silicate form the relatively insoluble calcium carbonate or silicate and thus remove the Ca^{++} from solution.

Theory of deflocculation. If a suspension of monodisperse hydrogen clay particles is allowed to settle for a long time, the particles will collect in a layer at the bottom of the container. When this equilibrium condition is reached, there will be a balance between repulsion forces, attractive forces, and gravity forces.

If the settled clay layer has the water decanted off the top and is dried very slowly, it is possible to measure the volume and calculate the water film thickness as previously explained. This thickness is found to be greater than that in plastic clay, since the forces are less. However, the bulk density of the dried piece will be lower than that of a corresponding piece dried from a plastic mass since there has not been an opportunity for the particles to pack closely.

Furthermore, the suspension discussed above, when examined under the microscope, is seen to be composed not of individual particles, but of small flocks of many particles. For this reason the suspension settles rapidly and leaves a sediment of low density.

Contrast this with a similar kaolinite fraction that has been deflocculated with NaOH. Here the particles are individuals that move about by Brownian movement; therefore, the settling is very slow. If sufficient time is given, however, the sediment is more dense than for the previous case, since each crystal fits into a closely packed arrangement. This at times produces an almost rocklike sediment.

From this simple evidence it may be deduced that in the flocculated system there are between the particles attractive forces that draw them together into flocks; however, at a certain distance the repulsion forces increase to balance them and maintain a stable condition. On the other hand, in deflocculated suspensions there is no evidence of any attractive force. This is confirmed by the fact that flocculated systems have definite yield points while deflocculated systems do not.

This force system may be represented with a reasonable degree of certainty in the diagram of Fig. 10–4. Here are plotted the repulsion forces between particles, which are quite precisely known, and the attractive and total forces *a*, which are only estimated. It will be noted that the force between particles is zero at a spacing (water film) of .04 micron. This is a point of equilibrium because the slope of the total force line passing through the axis is negative. That is, if the particles are slightly separated beyond this distance, the forces built up are attraction forces and tend to bring them together again. On the other hand, if the particles are forced together, repulsion forces are built up to push them apart. This is a simple, clear-cut picture and seems to explain the behavior of clay suspensions.

In the case of a deflocculated suspension, the total force curve is represented by *b*. Since there are no attractive forces, there can be no yield point. Curve *c* is the total force for a flocculated system at higher pressure (curve *a* moved upward) whereby the stability point is moved to the left. This picture is complicated by the possibility of polarization of the clay plates, that is, some areas with positive charges and some with negative charges.. This condition would aid in forming scaffold-like flocks.

Mechanism of deflocculation. We have discussed what happens in deflocculation, but to explain why it happens is not at all simple. The theory most generally accepted by the colloid chemists is briefly described below, but it should be remembered

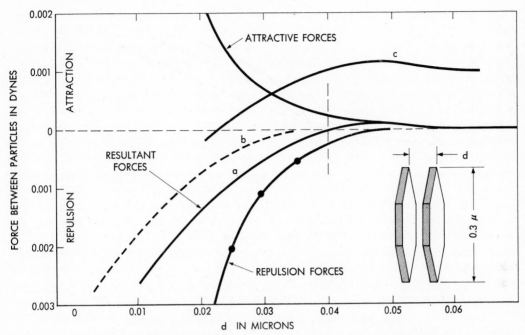

FIG. 10–4. The forces between a pair of kaolinite plates in suspension.

that at present it is only a theory and there is little proof to substantiate it. In Fig. 10–1 there was shown a section of the kaolinite structure with an edge having broken bonds. It is supposed that $(OH)^-$ are preferentially adsorbed at the positive points in the lattice structure. In the hydrogen clay H^+ act as counter ions in the surrounding water hull. The size of this hull changes with the type of counter ions, becoming larger with the more highly hydrated Na^+. As the NaOH concentration increases, the negative charge on the particle increases, as was shown in Fig. 10–2, because the increased Na^+ outside the water hull will leave some $(OH)^-$ unneutralized. The hydrogen clay should have no charge and a pH of 7, but this is not exactly realized because of dissolved impurities in the distilled water. Adding NaOH up to the base exchange capacity makes no great change in pH because of a sort of buffer action, but as soon as the base exchange capacity is exceeded the Na^+ and $(OH)^-$ will exist in the free water medium and will prevent the building up of any attractive forces.

Casting process

Essentially the casting process consists of pouring a slip into a dry plaster mold which absorbs water from the slip until a layer becomes rigid enough to support itself.

Process of water absorption. As a simple case, take that of a dry plaster mold against which is suddenly poured a clay slip, shown sectionally in Fig. 10–5. At the instant of pouring assume that the slip contains 20 per cent of water (based on dry weight) and that the plaster is dry. After an interval of time, the slip next to the plaster will have lost water and the plaster will have gained water. At later times the transfer of water

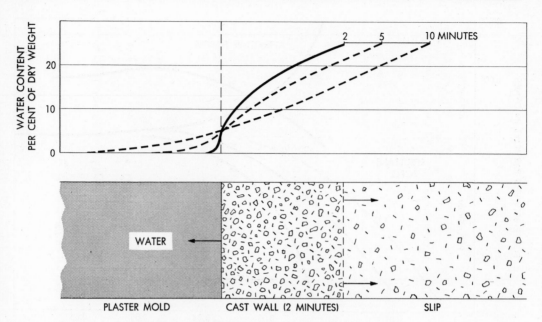

FIG. 10–5. Schematic diagram of the casting process.

will continue as shown in the diagram. The water flowing into the plaster must pass through the stiffened wall of clay adjacent to the plaster, which acts as a filter. Therefore, the rate of casting depends largely on the permeability of the clay layer. For example, bodies high in ball clays cast slowly, while those containing coarse clays cast rapidly.

Purpose of deflocculating casting slips. There are a number of reasons why slips are deflocculated for casting purposes. The most important one is the fact that a firm layer of clay is formed next to the plaster, with a sudden gradation to the liquid slip. Another reason for deflocculating is to reduce the water content of the cast layer. Also, deflocculated slips have little tendency to settle, so a uniform wall thickness from the top to the bottom of a cast is obtained.

Preferred orientation. As indicated earlier, the clay particles are platelike, and it

might be expected that in the casting process they would lay down in an orderly manner, much like shingles on a roof. Indeed, there is good evidence that this is the case, since a layer cast against plaster has a greater drying shrinkage in its thickness than in its length or width. In other words, there are more films per unit distance normal to the plaster surface than in other directions.

The very interesting photograph in Fig. 10–6 shows the preferred orientation on the fractured surface of a dried cast. The growth lines are very much like the dendritic crystals produced in the solidification of a metal casting.

Segregation. As most casting slips are made up of clays and nonplastics, it might be suspected that the wall of the cast would not be homogeneous. There has been little study given to this problem, but the evidence points to good uniformity in the cast

FIG. 10–6.   Broken section of a porcelain cast, showing the method of building up the wall.

layer.   The water content, however, is not uniform from face to face, as shown in Fig. 10–5, and this causes a differential shrinkage in drying and even in firing.   For example, if a thin plate is cast on a single surface mold, the edges will curl away from the plaster on drying because of a differential shrinkage through the layer.   If the piece is cylindrical like most pottery, the warpage cannot occur; nevertheless, strains are set up.   A flat tile may be cast in a flat plane in a two-surface mold, since here the moisture gradient is symmetrical.

Control of casting slips.   The building up of a good casting slip, it will have to be admitted, is at present as much an art as it is a science.   Until we have more fundamental knowledge of the subject, this condition is going to exist.   However, it is possible to apply science to some of the problems, and these cases will be pointed out.

Properties of casting slips.   There are certain desirable properties of the casting slip, many of which can be measured quantitatively to give an excellent over-all picture.   These are:

1. A low enough viscosity to flow into the mold readily.
2. A low rate of settling out on standing.
3. Ability to drain cleanly (in drain casting).
4. Giving sound casts in solid castings.
5. Stability of properties when stored.
6. Quick release from mold.
7. Proper casting rate for each operation.
8. Low drying shrinkage after casting.
9. High dry strength after casting.
10. High extensibility when partly dried.
11. Freedom from trapped air.
12. Freedom from scumming.

Types of material.   The usual whiteware casting slip is composed of kaolin, ball clay, Florida kaolin, and the nonplastic.   The

Table 10-1

Composition of Some Casting Slips for Whitewares

Constituent	Sanitary ware, drain cast (per cent)	Sanitary ware, solid cast (per cent)	Hotel china (per cent)	Semivitreous tableware (per cent)	High-tension electrical insulators (per cent)
Feldspar	32	34	25	14	33
Flint	20	17	37	35	21
Washed Georgia kaolin	22	24	21	28	23
Washed Florida kaolin	6	9	7.5	9	8
Ball clay	20	16	8	15	15
Whiting	--	--	1.5	--	--

ball clays slow the casting rate and if used in too large an amount (over 22 per cent) are likely to cause soft spots. They also increase the green strength of the ware. The kaolins give faster casting but lower the green strength.

Some casting slip compositions for whiteware bodies are given in Table 10–1.

The composition of the slip may well be studied from the point of view of particle packing to get the most dense structure.

Deflocculants. It was shown previously that natural clays containing adsorbed divalent ions could not be deflocculated with a simple base like sodium hydroxide. However, sodium carbonate and sodium silicate will work and are generally used together in the industry. The proportions of the two will depend on the clays in the body. The leaner clays work better with a higher silicate of soda content, since this produces some colloidal silica. Organic deflocculants, such as humic acid, have been used to some extent. Also, ammonium sulphonates have been used industrially in some casting slips.

Nonclay casting slips

It is possible to make excellent casting slips from pure fused oxides, nitrides, and

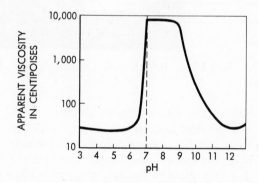

FIG. 10–7. The apparent viscosity of fused alumina particles of 5 micron diameter suspended in water.

other nonplastics. The material is ground down to an average particle size of 5 microns, treated with hydrochloric acid to lower the *pH*, and adjusted to the highest possible specific gravity. The viscosity curves for alumina treated in this way are shown in Fig. 10–7. This amphoteric substance may be deflocculated either with an acid or a base. Some materials, such as magnesia, would be hydrated in water, and thus are suspended in absolute alcohol for casting.

REFERENCES

HALL, F. P., "The Casting of Clay Ware: A Résumé." *J. Am. Ceram. Soc.* **13,** 751, 1930.

HAUSER, E. A., *Colloidal Phenomena.* McGraw-Hill Book Co., Inc., New York, 1939.

JOHNSON, A. L., and NORTON, F. H., "Fundamental Study of Clay: II, Mechanism of Deflocculation in the Clay-water System." *J. Am. Ceram. Soc.* **24,** 189, 1941.

CHAPTER 11

FORMING METHODS

Introduction

The methods of forming ceramic ware are generally divided into four classes based on the consistency of the mixture. These are the dry press method, the extrusion method, soft mud molding, and casting. Each of these will be discussed in this chapter.

Dry press method

This method is used for making small electrical insulators, tiles, and refractories. The water content of the pressing mixture is low — 5 to 15 per cent, and the pressure is high — several thousand lb per sq in.

Preparation of the mix. The dry press mix for whitewares is generally prepared by passing the partially dried filter cakes through a dust blower and then through a screen. On the other hand, some potteries have recently employed the dry mixing method with air-floated clays. Grog mixes are generally prepared in a wet pan or special mixer.

Pressure uniformity. One of the problems in dry pressing is to attain a uniform density throughout the die. Taking as an example a simple cylindrical die, the density distribution will be somewhat as shown in Fig. 11–1(a) if the pressing is from one side. On the other hand, if the pressure comes from both top and bottom as in Fig. 11–1(b), the density uniformity is greatly improved. However, if a lubricant is used on the die walls or in the mix, a still greater improvement occurs, as shown in Fig. 11–1(c). For

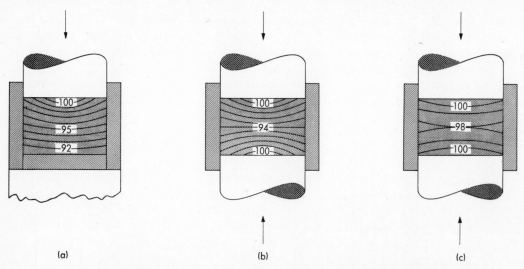

(a)   (b)   (c)

Fɪɢ. 11–1.   Density distribution in a die under various pressing conditions: (a) pressed from one side; (b) pressed from both sides; (c) pressed from both sides with a lubricant.

this reason, small pieces often contain a lubricant, especially if they are low in the natural lubricants, clay and talc. Also, most pressing is done from both sides, or else a floating mold box is used to obtain the same condition.

Magnitude of pressure. The pressure in use runs from a few hundred lb per sq in. for high clay bodies to as much as 100,000 lb per sq in. for special refractories. In most cases it is desirable to obtain the greatest possible density compatible with mold wear; however, there is an upper limit to which the pressure can be taken in a production cycle, since air is trapped in the pores with no time to escape. Upon release of the pressure, this highly compressed air expands and opens up cracks around the sides of the piece that are called pressure laminations.

There are two ways in which this cracking may be minimized, and higher working pressures permitted. One of these, commonly used in Europe, is to press at first to just below the cracking pressure, release this pressure for a short time to let the air escape, and then re-press at a much higher pressure. This method works quite well but reduces the capacity of a given press.

The other method consists of evacuating the air from the mix by connecting the die box, just after the plunger has entered, with a vacuum tank. This method is used to a considerable extent for large pieces, such as heavy refractories.

Mixtures for dry pressing. The mixtures for pressing whitewares are similar to those for plastic molding. However, high talc bodies which might well lack the plasticity for other forming methods may be easily pressed. Also, completely nonplastic bodies, such as fused oxides, may be pressed when proper plasticizers are used. In the refractories industry, bodies that contain a

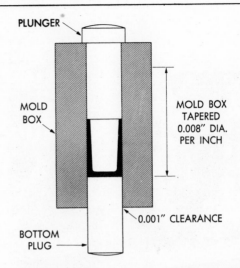

FIG. 11–2. Example of a crucible die. All working parts should be highly polished and the chrome-carbon die steel should be heat treated to give a hardness of Rockwell C, 58–62.

large proportion of coarse grog are pressed quite successfully in hard dies.

Lubricants and plasticizers. Organic compounds, such as starches, corn products, and waxes, may be added to the mix to increase the plasticity or flowability and to reduce the friction on the die walls.

Die design. The design of dies requires much experience, and it is not a subject that can be treated to any great extent here. However, as an example, a small crucible die with workable tapers and clearances is shown in Fig. 11–2. A highly polished surface is always desirable for the inner walls.

Production dies are made of cast iron, chromium-plated mild steel, hardened steel, stellite, or tungsten carbide, depending on the abrasive qualities of the mix.

Ramming methods. When making large refractory pieces, it is not economical to press them as previously described, since the die cost would be extremely high, and uniformity of pressure would be very diffi-

FIG. 11–3. Photograph of a quick acting hydraulic press (Dennison Engineering Company).

cult to obtain throughout the large mass. Under these conditions, it has been found better to ram the mix, little by little, into the mold with a mallet or air hammer. Care must be taken to consolidate each addition with the previous one to prevent laminations.

Presses. A large number of types of presses are used in the industry. Small pieces were at one time made on screw presses where an inertia weight gave an impact blow. Today most modern plants use quick acting hydraulic presses, such as that shown in Fig. 11–3 for tile and electrical insulators.

Some refractories are made on hydraulic presses, but a heavy toggle press is generally used because of its fast action. It is expected, however, that here, too, the specially designed hydraulic press may eventually supersede it.

Extrusion method of molding

This method employs the body in the form of a plastic, but very stiff, paste. In general, this mix is forced through a die to form a continuous column which may then be cut into appropriate lengths.

Piston extrusion. Some extruders, for example the sewer pipe press, force the mix through the die by means of a piston energized by steam or air pressure. This is, however, an intermittent process, little used in the ceramic industry.

Augers. The usual extrusion machine is called an auger because the feeding screw has this appearance. A cross section of a typical auger is shown in Fig. 9–6. The die may be of any shape that produces the desired column, even if the cross section is quite complicated, as shown in Fig. 11–4. In case the mix is abrasive, the auger and

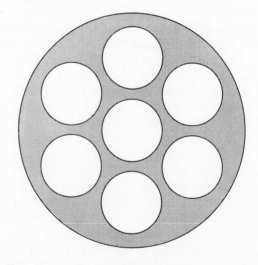

FIG. 11–4.   Cross section of a column extruded from an auger.

the die may be made of a wear-resisting metal like stellite. Much experience is needed to design the die to produce a smooth, uniform column. In many cases, the die is heated and lubricated to reduce friction.

As stated in Chapter 9, many clays are improved in working properties by de-airing; thus vacuum augers are now commonly used in the industry. The column is more homogeneous and dense after this vacuum treatment.

Extrusion of nonplastics. Completely nonplastic bodies, if finely ground, may be extruded through a die, provided they contain sufficient plasticizer.

Injection molding. This process has been highly developed for the high speed molding of exact shapes in the organic plastics industry. It is possible to use the same method to force a finely ground nonplastic, mixed with about 15 per cent of thermosetting and thermoplastic resin, from a heated chamber into a cooled mold. The chilled piece is heat-treated to remove the organic matter and is then ready for the firing operation. This method is used for small pieces, such as sintered alumina cores for spark plugs.

Soft plastic molding

This is the earliest method of forming clay. It may be accomplished with the hands alone in building vessels with coils, as is still done by primitive people, or in throwing on the potter's wheel. Also, early types of brick were handmade in wooden molds from a soft plastic mix. Today little but art pottery is handmade in this country, but the method is used in a number of production processes.

Jiggering process. This process is used largely in the whiteware industry to form plates and some types of hollow ware. Un-

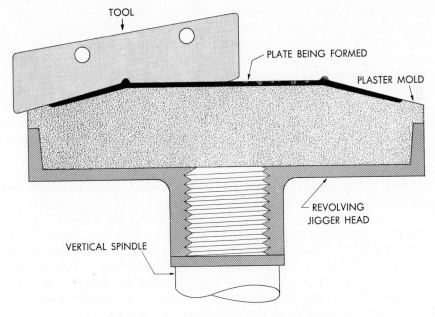

FIG. 11–5. Cross section of a jigger head and plate being formed.

til recently, hand jiggering was the rule, but now many potteries use automatic jiggers for everything except a few specialties.

The operation starts out with a lump of plastic body of the proper weight. This is then formed into a round bat like a pancake by striking with a plaster tool or by spreading it on a revolving disk with a descending tool. As this is the only part of the process where the body is appreciably deformed, it is vital that the finished bat be completely homogeneous. The bat is next transferred to a plaster mold shaped like the upper surface of a plate, for the plate is formed upside down, as shown in Fig. 11–5. This mold is then set in a chuck on the upper end of a vertical shaft that is turning at the rate of 300 to 400 rpm. A tool contoured like the underside of the plate is then pulled down to make contact with the bat, now lubricated with water, and forms the surface precisely, partly by scraping off excess body and partly by forcing the body down on the mold. A section of the tool in Fig. 11–6 shows this operation. The plaster mold and plate are now taken from the jigger chuck and sent through a continuous dryer. A typical hand jigger operation is shown in Fig. 11–7; here a crew of three produces around 40 dozen plates per hour.

In cases where the body is so low in plasticity that handling the bat is impossible, as happens with bone china and frit porcelain, the bat may be formed on a cloth-covered ring for transfer to the mold, or the bat may be slip-cast onto the jigger mold and then formed.

The jiggering operation requires great skill to form large plates free from warpage, for any unequal strain at any stage of the process will show up in the firing. There is also a considerable amount of preferred orientation of the clay particles on the jig-

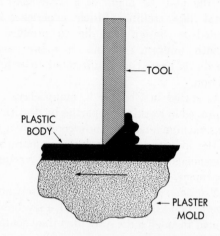

FIG. 11–6. Operation of the jigger tool.

gered surface, which contributes to the smooth finish attained in this operation.

It was long thought that such an exacting operation could not be done by machine, but as early as the year 1935 an automatic jigger for cups was in use in Sweden. About the same time an automatic jigger of great capacity was developed by the Homer-Laughlin Company of Newell, West Virginia, for flat ware. During this period W. J. Miller developed and put on the market an automatic jigger that is now used in many potteries.

The Miller jigger operates as shown in Fig. 11–8. A round column of de-aired plastic body is fed to the machine, where it is cut off in correct lengths and dropped on the jigger mold. This mold is then pressed up against a heated die that spreads it on the plaster mold almost to its final size. The die is heated so that the resulting film of steam will prevent sticking of the body. The mold and formed body are then placed on the jigger head where they are held by vacuum. The jiggering operation takes off little clay, but produces a smooth finish as

FIG. 11–7. Hand jiggering operation (Homer-Laughlin Company).

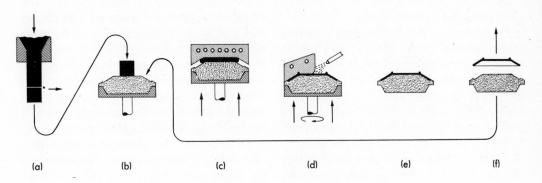

(a)   (b)   (c)   (d)   (e)   (f)

FIG. 11–8. Operation sequence in the Miller jigger.

the surface is sprayed with a water mist. The jiggering speed is much higher than in hand operation, 500 to 1200 rpm. The mold and finished plate are then carried to a continuous dryer, where the plate separates from the mold. It is then removed and the edges are trimmed. The capacity of an eight-line machine is about 600 dozen per hour; consequently it saves a great amount of labor compared with hand operation.

Hand molding. There are some products, such as face brick and special refractory shapes, that are hand molded from plastic clay. A lump of clay of the correct size is thrown into a mold, and the excess is cut off with a wire. Also, the larger terra cotta shapes are hand molded in plaster molds.

Plastic pressing. Recently, plastic masses have been pressed between porous, hard plaster molds, with means to blow or suck excess water from the pores at each cycle.

Slip casting

This operation is used to form many different articles. It is applied to all pottery not having the shape of a surface of revolution that would adapt it to jiggering, and includes such forms as vegetable dishes, teapots, or artware. It is also used for heavy ware in the plumbing industry, for large glass pots, and for tank blocks.

Plaster molds. Since all casting is carried out in plaster molds, some space will be devoted to the nature of plaster of Paris. The raw material, as described in Chapter 24, is gypsum rock with the composition $CaSO_4 \cdot 2H_2O$. This rock is finely ground and calcined to produce the hemihydrate $2CaSO_4 \cdot H_2O$, which is the plaster used by the potter. When water is added to this plaster, it recrystallizes to form again the $CaSO_4 \cdot 2H_2O$ with considerable strength.

The properties of the plaster may be varied by the amount of water added, as shown in the diagram of Fig. 24–5.

Hand mixing of plaster and water has been largely superseded by mechanical stirrers. Typical mixes used in the pottery are:

Casting molds —
　　　36 lb of water plus 40 lb of plaster
Jigger molds —
　　　39 lb of water plus 50 lb of plaster

The plaster is soaked for 2 minutes and stirred for $3\frac{1}{2}$ minutes. There are special plasters of greater strength made by different calcination methods. They are much used now for models and mold cases.

The art of mold making will not be discussed here, since it can only be learned by doing it in a laboratory course. However, it might be well to say something about parting solutions that prevent cast plaster from sticking to another plaster surface. There are many kinds used by sculptural casters, such as shellac, stearine in kerosene, beeswax in benzol, and silicate of soda. On the other hand, the potter uses soap almost exclusively. This is a high fat soap known as English Crown Soap and sold by potters' supply houses. A thick solution of the soap is made up and sponged on the damp plaster surface until a good lather is built up. After a few minutes this is wiped off with the rinsed-out sponge and a second coating put on which is again sponged off. It is important that all suds be completely removed. If the soaping is properly done, the plaster surface will have a dull sheen like ivory, and when dry can be polished. The soap forms with the plaster a calcium oleate which is insoluble and fills the surface pores. Soap is preferred by the potter over other partants, since it does not affect the surface of the piece cast against it as oily or waxy partants would.

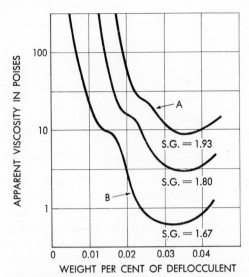

FIG. 11–9. Viscosity curves for a whiteware slip. Solid castings would be made at (a) and drain castings at (b).

Table 11-1

Grog Casting Slip for Glass Pots[1]

North Carolina kaolin	13
Georgia kaolin	9
Tennessee no. 5 ball clay	10
Kentucky no. 4 ball clay	10
Maine feldspar	5
Grog: 10 - 20 mesh	18
20 - 40 mesh	16
40 - 100 mesh	10
through 100 mesh	9
Silicate of soda ("S" brand)	0.025
Sodium carbonate	0.012

[1] From Heindl

Commercial casting slips. The composition of typical whiteware slips has been given in Table 10–1. Good casting slips may also be made up containing talc, rutile, and other materials used in special electrical porcelains.

The deflocculation of the slip was formerly an art, but it now can be carried out scientifically. This is done by making up a series of slips with varying specific gravities and varying amounts of deflocculant. At first the latter may be one-half silicate of soda and one-half sodium carbonate, but other ratios may be tried later.

After aging and agitating to reach equilibrium, each of the slips is tested for apparent viscosity. Of course, with experience the number of samples may be reduced. In Fig. 11–9 a set of viscosity curves made in this way is shown. It will be noted that there is a double minimum which has been explained as due to the different base exchange capacity of the kaolin and the ball

clay; this explanation, however, may not be the true one. It is well to keep the amount of deflocculant just below the second minimum on the curve, since any over-deflocculation causes sticking to the mold and often ruins the mold by "burning" the surface, that is, forming a layer of insoluble calcium silicate in the surface pores of the plaster. In Fig. 11–9, a drain casting slip might be deflocculated at *B* and a solid casting slip at *A*. In general, it is desirable to keep the specific gravity of the slip as high as possible and still have it fluid enough to pour.

Casting slips may be made from coarse grog for forming refractories like glass pots or blocks. A typical slip given by Heindl is shown in Table 11–1.

Nonplastic casting slips. As stated in the last chapter, a nonplastic may be slip cast under certain conditions. In Table 11–2 are the slip characteristics for a number of fused oxides that may be readily cast.

Drain casting method. In this method the cast is made from one surface and is therefore especially adapted to thin ware, as shown in Fig. 11–10. It is not suitable for

Table 11-2

Pure Oxide Casting Slips

Material	Time of milling in hours	Specific gravity of casting slip	pH	Raw material
BeO	18[1]	2.0	4.0	Brush el. fused - 325 m
Al_2O_3	24[1]	2.4	3.4	Fused - 325 m
ZrO_2	24[1]	3.3	2.3	T.A.M. fused, stab. - 325 m
MgO	20[2]	2.5	--	Norton Co. fused - 325 m

[1]Gallon steel mill, dry with 1200-1800 gm charge. Mill turns 58 rpm. Acid washing removes iron.
[2]Gallon porcelain mill with absolute alcohol.

slips containing coarse grog, since in this case the inner, drained surface would be rough.

One of the troubles encountered in casting is a slight settling of the slip which causes a heavier wall thickness at the bottom of the piece. This is caused by too low a specific gravity of the slip or by improper deflocculation. Another defect is that of pinholes caused by air bubbles in the slip. Poor draining is caused by the wrong mixture of clays in the body or by improper deflocculation. Still another defect is called wreathing, wherein fine lines appear around the piece because the slip rises in little jumps up the mold surface. Fast pouring, or vibration of the mold will help this condition.

Solid castings. A typical solid cast mold is shown in Fig. 11–11. Anyone familiar

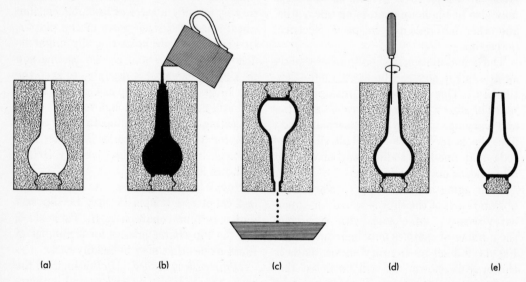

(a)   (b)   (c)   (d)   (e)

Fig. 11–10. Operations of drain casting: (a) assembled mold; (b) pouring slip and casting; (c) draining; (d) trimming the top; (e) disassembling mold and removing finished piece.

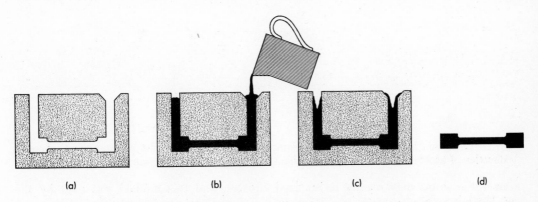

Fig. 11–11. Operations of solid casting: (a) assembled mold; (b) filling the mold; (c) absorbing water from the slip; (d) finished piece, removed from the mold and trimmed.

with foundry practice will at once see the similarity between this and metal casting. There must be inlets for entrance of the slip, vents for the air to escape, and a reservoir of slip to supply the casting as it shrinks.

The casting proceeds from all sides simultaneously, so that the walls grow until they meet in the center. It is essential that means be available to supply this liquid core with additional slip until the piece is solid. Otherwise a hollow will result, for even the best slips have a considerable volume shrinkage in casting. In general, the properties of the slip for solid casting are more critical than those for drain casting. Very large sections may be solid cast, for example tank blocks 12 to 18 in. thick. In this case the slip has a high grog content and is so well deflocculated that it contains not over 12 per cent of water.

Scrap problems. As the scrap from draining and trimming cast pieces contains some deflocculant as well as some calcium sulphate from the mold, great care must be taken to prevent unbalance of the deflocculation when returning this material to the blungers. In some plants the scrap is washed to remove all soluble salts, and is filter pressed and added back as raw material.

Electrolytic casting. As clay particles carry a negative charge, they may be deposited on a metal mold that acts as an anode. Moreover, a slip of a normal whiteware composition may be deposited in the same way, producing excellent detail. Because of the difficulty of controlling the wall thickness, this method is not used commercially.

Finishing ware

The finishing is an important step in making high grade ware. A few of the more common processes will be outlined here.

Trimming. Jiggered ware must have the feather edge on the rim trimmed off with a scraper and then sponged to make it smooth. Cast ware must have the top trimmed off, that is, the extra part that holds the slip for shrinkage, commonly called the spare. This trimming is usually done in the mold as shown in Fig. 11–10.

In many pieces there are seams left by the joints in the mold. In the case of earthen-

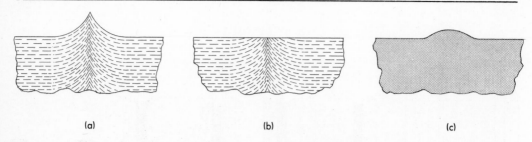

FIG. 11–12. Seam left when casting porcelain: (a) cross section of cast piece at seam, showing direction of kaolinite plates; (b) same section sandpapered smooth; (c) same section after firing.

ware these seams may readily be sponged off, but not so in vitreous ware. Here, no matter how perfectly they are leveled off in the dry state, they appear again after firing. This is a very interesting phenomenon which may be explained by the preferred orientation of the clay particles during casting somewhat as shown in Fig. 11–12. Here it is seen that particles at the seam are normal to the general surface, while all the others are parallel to it. Therefore, the thickness shrinkage in firing will be greater for the average area than at the seam, which thus is left protruding. This is an excellent example of orientation influencing shrinkage. There are a number of ways of getting around this trouble. One method used with bone china consists of hammering the freshly cast piece with a small hammer at and about the seam to cause plastic flow and give a random orientation. Another method used with high fire porcelain is to fire just below the maturing temperature, grind the seam down with an abrasive wheel and then refire.

Turning. Some fine ware, after forming and while partially dried, in what is often called the leather hard stage, is placed in a chuck on a lathe or on a potter's wheel and turned with steel tools to exact dimensions. The feet of cups are often turned in this way. The eggshell porcelain of the Chinese was made by turning from heavier thrown pieces, and this method was used by the ancient Greeks to form their exquisite vases. Large high-tension insulators are now often turned out of a dried blank on a lathe. As nearly all drying shrinkage has taken place before turning, very precise shapes may be arrived at in this way.

Burnishing. The glossy appearance of the ware made by our southern Indians is due not to a glaze but to a burnishing operation, in which a polished pebble is rubbed over the surface of the leather hard piece. The same method may be carried out more expeditiously in the lathe, for example, in the case of Wedgwood jasper ware. Here again the mechanism of burnishing is the production of a surface layer of clay plates all laid down parallel to the surface like shingles on a roof. Even the firing operation does not entirely destroy the polish.

Jointing or sticking up. Much ceramic ware is fabricated in pieces and then joined together by using slip as glue. Examples are the handles on cups or a complex piece of sanitary ware. Some of the porcelain figure groups are assembled from as many as sixty pieces. The edges to be joined are roughened and then coated with slip and quickly assembled. The important thing is that the two pieces to be joined have exactly the same water content. If this condition is observed, there will be no differential shrinkage in drying.

REFERENCES

DODD, C. M.; *The Effect of Various Factors on Pressure Transmission in Dry Pressing.* Am. Ref. Inst., Tech. Bull. No. 33, 1932.

EVERHART, J. O., AUSTIN, C. R., and RUECKEL, W. C., *The Effect of De-airing Stiff-mud Bodies for Clay Products Manufacture.* Ohio State Univ. Studies, Vol. 1, No. 6, 1932.

GOULD, R. E., *Making True Porcelain Dinnerware.* Industrial Publications, Inc., Chicago, 1947.

HALL, F. R., "Whiteware," from *Encyclopedia of Chemical Technology.* Inter-science Encyclopedia, Inc., New York, 1949, Vol. 3, p. 545.

HAUTH, W. E., JR., "Slip Casting of Aluminum Oxide." *J. Am. Ceram. Soc.* **32,** 394, 1949.

HEINDL, R. A., MASSENGALE, G. B., and COSSETTE, L. G., "The Slip Casting of Clay Pots for the Manufacture of Optical Glass at the National Bureau of Standards." *Glass Industry* **27,** 177, 1946.

NEWCOMB, R., JR., *Ceramic Whitewares.* Pitman Publishing Corp., New York, 1947.

CHAPTER 12

DRYING OF CERAMIC WARE

Introduction

The drying process is an important step in the manufacture of many ceramic articles. While the dictates of economy require the fastest possible drying, too fast a schedule causes differential shrinkage of such magnitude as to produce cracking. In this chapter the principles of the drying of porous solids will be discussed.

It is assumed that the student is already familiar with the properties of air, and has some knowledge of psychrometry; if not, this knowledge may be obtained from the references at the end of the chapter. However, it should be kept in mind that moving air serves a twofold purpose in the drying process; it supplies heat to the ware as compensation for the evaporative cooling, and it carries away the water vapor formed.

Internal flow of moisture

Water evaporated from a piece of ware by drying must come mainly from the interior of the piece through the fine inter-connecting channels.

Internal flow. The rate at which this water flows through a given structure, as shown in Fig. 12–1, is given by

$$\text{Volume rate of flow} = k \frac{\text{driving force}}{\text{flow resistance}},$$

or

$$\frac{dV}{dt} = \frac{k(C_2^1 - C_1^1)}{l} \cdot \frac{p}{\eta},$$

where dV/dt is the volume rate of flow, C_1^1 is the water concentration on the wetter face, C_2^1 is the water concentration on the dryer face, k is a constant, l is the length of path, p is the permeability of the body, and η is the viscosity of water.

From the above relation it is evident that

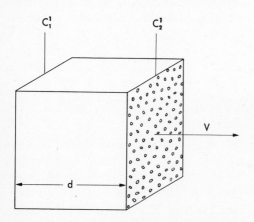

FIG. 12–1. Movement of moisture through a porous medium with a moisture of gradient of $(C_1^1 - C_2^1)/d$.

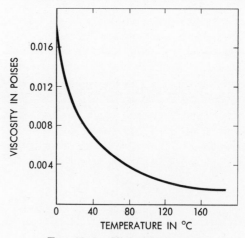

FIG. 12–2. Viscosity of water.

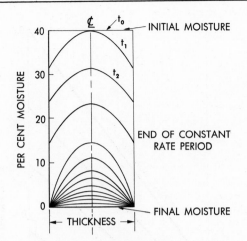

FIG. 12–3. Moisture distribution in a slab of
drying clay.

to increase the rate of water flow in a given
material we have the choice of increasing
the permeability, increasing the moisture
gradient, or decreasing the viscosity of the
water. Increasing the moisture gradient,
as will be shown later, cannot be carried beyond a certain point without causing rup-
ture of the body; on the other hand, the
viscosity of the water may be decreased by
working at higher temperatures, as shown
in Fig. 12–2, and the permeability increased
by a coarser structure.

Moisture distribution. Many studies have
been made of the moisture distribution in a
drying solid. Figure 12–3 gives a typical
distribution for a drying slab with no loss
from the edges. The lines of equal moisture content at first are nearly flat but soon
become highly curved. The moisture gradient is, of course, given by the slope of
these lines. The more slowly the drying
takes place the less is the curvature.

Surface evaporation

The rate of evaporation from the surface
of drying clay ware depends on many factors
which will be covered in this section.

Evaporation from a free water surface.
The evaporation rate from a free surface is
dependent on the air temperature, the air
velocity, the water content of the air, and
the water temperature. These factors are
well known and can be found in treatises on
drying, such as that of Lindsay. A simplified chart is shown in Fig. 12–4 to give the
reader some idea of these relations.

Drying rates of ceramic bodies. Should
the weight of a drying piece be plotted
against time, the result would be a smooth
curve without any great significance. However, if the rate of drying, or the slope of the
weight loss curve, is plotted against the
water content of the piece, a curve like that
in Fig. 12–5 results. As first pointed out by
Sherwood, the wet clay starting at point *A*
drys at a constant rate until *B* is reached.
This constant rate of drying is exactly the
same as for a free water surface. At point
B the rate starts to decrease rapidly and
finally reaches the origin. It is significant
that point *B* is where the mass changes from
a dark color to a light color. In other
words, down to point *B* there is over the
surface a continuous water film which acts
as free water, but below this point the water
retreats further and further into the pores
so that the drying rate becomes less and
less. These steps in the clay structure are
shown by Fig. 12–6. In the next section it
will be shown that this behavior is closely
connected with the drying shrinkage.

Drying shrinkage

Mechanism of drying shrinkage. A plastic ceramic body may be dried slowly under
conditions that permit a continuous measurement of the weight and the volume.
From this data it is then possible to construct a curve as shown in Fig. 12–7 which
tells an interesting story. The drying is
started at *A* with a uniform volume de-

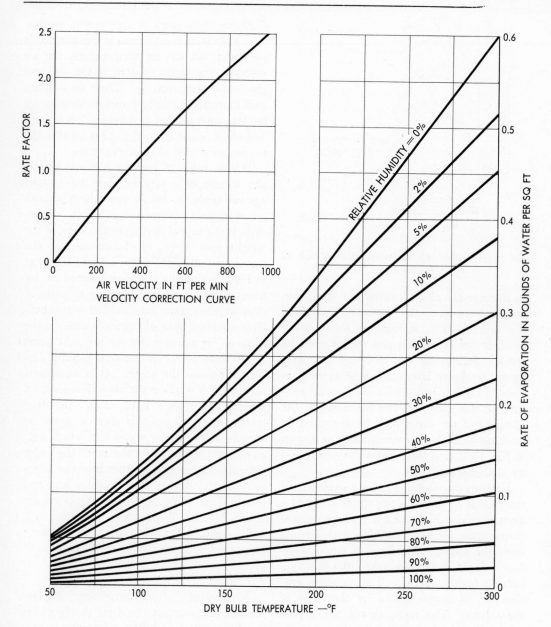

FIG. 12–4.   Evaporation rates from a free water surface.

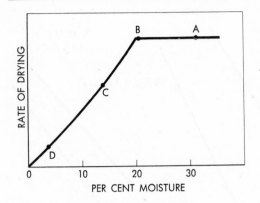

FIG. 12-5. Rate of loss in drying moist clay.

kaolin which shows a slight expansion at the very end of the drying period. There is no known explanation for this behavior. Curve 2 is for a ball clay which shows a slight secondary shrinkage as it reaches dryness. This is probably caused by a small amount of montmorillonite in the clay which retains water between the lattice planes until the very end.

The total shrinkage of a clay varies a great deal, the finer grain causing greater shrinkage. Table 3-4 gives shrinkage values for a number of clays.

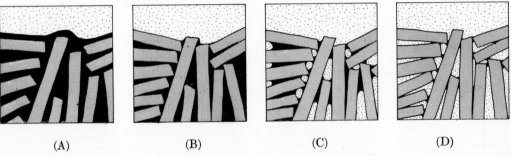

(A)   (B)   (C)   (D)

FIG. 12-6. Stages in drying moist clay. A cross section at the surface. The letters correspond to those in the curve of Fig. 12-5.

crease just equal to the water lost until point B is reached, after which no further volume change occurs. If line AB is extended, it will pass through the origin at 45°.

The significance of this curve is not hard to see. Between A and B the water lost comes from the layers between the particles, so that the latter come closer and closer together until they touch at B and can consolidate no further. From B to C the bulk volume does not change and the water removed comes from the pores. Point B in Fig. 12-7 corresponds to point A in Fig. 9-1.

Shrinkage curves for clays. Some clays depart from this curve to a slight extent, as shown in Fig. 12-8. Curve 1 is for a residual

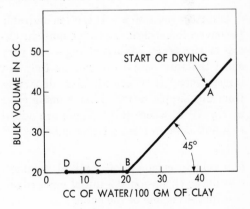

FIG. 12-7. Drying shrinkage curve of a moist clay. The letters correspond to those in Figs. 12-5 and 12-6.

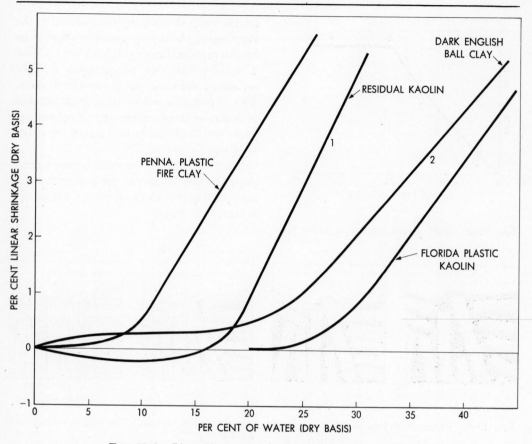

FIG. 12–8.   Linear drying shrinkage curves for several clays.

Shrinkage conversion.   It is often desirable to convert linear shrinkage to volume shrinkage or *vice versa*.   When changes are very small the ratio is 1 to 3, but as they get larger, there is a considerable departure from this simple ratio.   This is made clear in Fig. 12–9, where seven pieces are added to a unit cube to form a larger cube.

Let

b = volume expansion (% initial volume),
a = linear expansion (% initial cube edge),
1 = initial volume of the cube.

Then the final volume of the cube

$$= \frac{b}{100} + 1,$$

or, in terms of a, based on the sum of the volume of the seven pieces,

$$1 + 3\left(\frac{a}{100}\right) + 3\left(\frac{a}{100}\right)^2 + \left(\frac{a}{100}\right)^3$$
$$= \left(1 + \frac{a}{100}\right)^3,$$

as

$$\frac{b}{100} + 1 = \left(1 + \frac{a}{100}\right)^3;$$

then

$$\sqrt[3]{\frac{b}{100} + 1} = 1 + \frac{a}{100},$$

$$\frac{a}{100} = \sqrt[3]{\frac{b}{100} + 1} - 1,$$

or

$$a = 100\left[\left(\sqrt[3]{\frac{b}{100} + 1}\right) - 1\right].$$

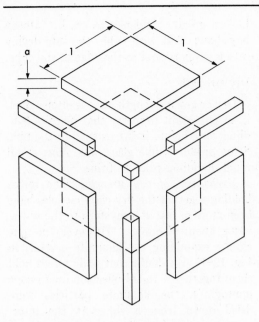

FIG. 12-9. The expansion of a cube.

In the same way it can be shown that to change linear shrinkage to volume shrinkage based on initial values the following relation applies:

$$b = 100 \left[\left(\frac{a}{100} - 1 \right)^3 + 1 \right].$$

It will be instructive if the student derives the two similar relations based on final rather than initial values.

There is much confusion about the base of the shrinkage calculation. For example, if a piece of plastic clay 10 mm long shrinks 1 mm, the linear shrinkage will be 10.0 per cent based on the initial length, but 11.0 per cent based on the final length. The base should be stated in all cases. In this book the base is initial length unless otherwise stated.

Methods of reducing drying shrinkage. Excessive shrinkage is undesirable as it tends to cause cracking and distortion of the ware. The commonest cure is to add nonplastics to the clay. These relatively coarse materials simply reduce the number of water films per unit distance by displacing a group of clay particles and their associated films by an equal volume of stable particles. In the same way coarse-grained clays shrink less than fine-grained ones. Because of the orientation of the clay plates, cast bodies shrink less (in the direction of the surface) than plastic ones.

Drying shrinkage can be greatly reduced by molding at high pressure so that the water films are reduced to a low order, as was described in Chapter 9. For example, many dry pressed bodies have an inappreciable amount of drying shrinkage.

Achievement of maximum drying rate

In any manufacturing operation, the greater the amount of salable product that can be turned out with a given piece of equipment, the lower will be the cost of the operation. This is true in drying, so every effort is made to dry ware at the maximum safe rate. Glass pots that formerly took months to dry may be dried with modern equipment in a few days.

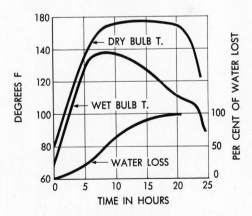

FIG. 12-10. Data for a typical dryer with humidity control.

Humidity dryers. It was explained earlier in this chapter that lowering the viscosity of water by high temperatures speeded up the drying operation. However, in most cases, putting the green ware into a high temperature would remove the water so fast that cracking would result. Therefore, the expedient of heating the ware all the way through in a saturated atmosphere was hit upon, for little water is lost from the piece under these conditions. Then, when the ware is thoroughly heated, the humidity is reduced as fast as is permissible without setting up dangerous stresses. A typical schedule for a humidity controlled firebrick dryer is shown in Fig. 12–10.

Size factors. In the case of several objects of different sizes from the same body, the larger ones will not only dry more slowly under a standard condition, but will also have a greater tendency to crack. Macey has shown that with cubes the safe drying rate is proportional to the edge of the cube.

Dry strength

Dry or green strength, as it is often called, is an important property that permits handling ware before it is hardened in the kiln. Some products, like glass pots, must stand rough handling prior to firing.

Mechanism of dry strength. The forces holding together the dry clay particles have been more or less of a mystery to those who write about ceramics. However, no one closely examining the kaolinite crystals in Fig. 12–11 can doubt that ionic forces hold them together. In this electron microscope photograph the kaolinite particles were dried down from a slip. As the interparticle water film disappeared, the forces

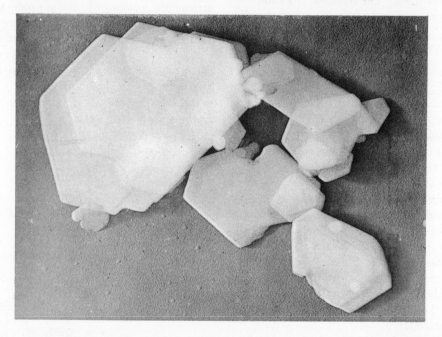

FIG. 12–11. Kaolinite crystals dried from a suspension to show similar orientations. The largest particle is 1 micron across. (From the thesis of Dr. Walter East. Taken by C. E. Hall, Massachusetts Institute of Technology, Cambridge, Massachusetts.)

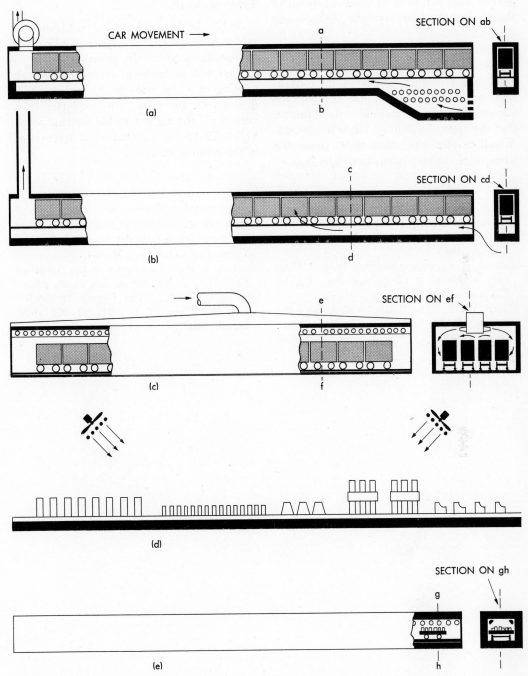

Fig. 12–12. Common types of dryers used in the ceramic industry: (a) steam-heated, continuous tunnel dryer; (b) waste-heated tunnel dryer; (c) continuous humidity controlled dryer; (d) hot floor; (e) radiant heat dryer.

present were sufficient to align nearly all of them so that their crystallographic axes were parallel.

Factors affecting dry strength. Clays vary greatly in dry strength, as shown in Table 3–5. The stronger clays probably contain some montmorillonite, while the weaker clays contain coarse particles. It is known that the type of adsorbed ion is important. A sodium clay has about three times the strength of the equivalent hydrogen clay.

Types of dryers

As this book is laid out to treat principles rather than design, little space can be given to specific dryers. In general, drying is carried out in modern dryers by hot air moving over the ware or by radiant heat directed on the ware. In the latter case, very rapid drying is possible for thin ware. Figure 12–12 shows a number of dryers used in the ceramic industry.

REFERENCES

GREAVES-WALKER, A. F., *Drying Ceramic Products*, 4th ed. Industrial Publications, Inc., Chicago, 1948.

LINDSAY, D. C., *Drying and Processing of Materials by Means of Conditioned Air*. Carrier Engineering Corp., 1929.

MACEY, H. H., "The Principles Underlying the Drying of Clay." *Trans. Brit. Ceram. Soc.* **33**, 92, 1934.

MACEY, H. H., "The Relative Safe Rates of Drying of Some Different Sizes and Shapes." *Trans. Brit. Ceram. Soc.* **38**, 464, 1939.

NORTON, F. H., "Some Notes on the Nature of Clay, II." *J. Am. Ceram. Soc.* **16**, 86, 1933.

SHERWOOD, T. K., "The Drying of Solids." *Ind. Eng. Chem.* **21**, 12, 1929; **21**, 976, 1929; **24**, 307, 1932.

CHAPTER 13

THERMOCHEMICAL REACTIONS

Introduction

Of all the steps in the process of producing ceramic articles, the firing is the most vital. Therefore we should learn all that we can about the reactions that take place at high temperatures. Of special interest is the equilibrium condition that is being approached, even though that condition is not always reached. Also, the rate at which the reaction progresses is often important.

Thermodynamics of reactions

Energy states. The stable phase at any temperature level is the one having the lowest free energy. For example, low quartz and quartz glass both exist at room temperature. If each is dissolved in hydrofluoric acid and the heat of solution meas-

ured, it will be found that the value for the glass is the higher. This shows that the glass has the higher free energy and is therefore the unstable phase.

Heats of formation. When two elements combine, they absorb or evolve energy ΔH, as shown for some of the oxides in Table 13-1. In general, if the heat energy evolved is large, the oxide has a high stability, as is the case for MgO or ZrO_2, but the fusion point is not exactly proportional to the heats of formation.

Vapor pressures. The stability of ceramic compounds is often indicated by the vapor pressure; the lower this value, the greater is the stability. Values for a few oxides are given in Table 13-2.

Table 13-1

Heats of Formation of Some Oxides

Oxide	ΔH^1 in cal/mol
MgO	-144,000
CaO	-152,000
Al_2O_3	-400,000
SiO_2	-206,000
ZrO_2	-259,000
BeO	-147,000
FeO	-65,000
NiO	-53,000
PbO	-53,000

[1]When ΔH is minus, an evolution of heat is indicated.

Table 13-2

Vapor Pressures of Some Oxides

Oxide	Pressure of O_2 in atmos at 1000^0K
CaO	4×10^{-56}
BeO	4×10^{-54}
MgO	3×10^{-53}
Al_2O_3	4×10^{-46}
ZrO_2	3×10^{-47}
SiO_2	4×10^{-36}
ZnO	2×10^{-28}
SnO_2	1×10^{-20}
CoO	9×10^{-18}
NiO	3×10^{-18}
PbO	2×10^{-13}
CuO	2×10^{-11}

Equilibrium conditions. The equilibrium between FeO and Fe_2O_3 is important in the color of glass. At high temperatures the equilibrium goes toward FeO and at higher oxygen pressures toward Fe_2O_3. The equilibrium under various conditions may be calculated from thermodynamic data. However, thermodynamics cannot compute the rate at which equilibrium is reached.

It is strongly recommended that the student of ceramics obtain a good background in that part of physical chemistry dealing with thermodynamic processes. There is space here for only a glance at the subject.

Phase rule

The statement of the phase rule by Gibbs was one of the great contributions to science. On it is based the physical chemistry of the ceramic field. Therefore, the student should be familiar with it as a groundwork for his ceramic work. There is no space here to do more than touch on this subject. It should have been covered by previous courses in physical chemistry.

The phase rule may be stated as

No. of phases + No. of degrees of freedom
$$= \text{No. of components} + 2,$$

or
$$P + V = C + 2. \qquad (13\text{--}1)$$

A phase is a physically homogeneous but mechanically separable portion of a system. Examples of phases in a system are ice and water.

The degrees of freedom of a system are the number of independent variables, such as temperature, pressure, and concentration, which must be fixed to define the system completely.

The components are the smallest number of independently variable constituents by means of which the composition of each phase can be quantitatively expressed.

For example, in the system Mg and O the components could be Mg, O; MgO, O; or MgO, Mg.

Equilibrium diagrams

The equilibrium diagram is a graphic representation of the phase rule as applied to a particular system. The two variables are usually temperature and composition. These diagrams give a complete picture of the system under specific conditions and thus are most useful for reference.

Method of making equilibrium diagrams. In the case of many metals the diagram may be readily determined by heating the components together in definite proportions until only the liquid phase is present. Then this liquid is allowed to cool slowly with a careful measurement of its temperature. The cooling curve will have breaks at the start and end of crystallization that make it possible to construct the diagram of this one composition readily. By repeating with other compositions the whole diagram may be obtained.

In the case of ceramic materials the heat evolved in crystallization is often small and the possibility of undercooling is great, so another method must be found. This consists of heating the intermittently mixed components to a given temperature until equilibrium is obtained and then quenching them to room temperature. A petrographic or x-ray examination will then give the kind and amount of phases present. This is repeated for other compositions and other temperatures until the diagram is completed. The complicated nature of the diagram, as well as the large number of runs needed, limits most work to two or three components.

Interpretation of equilibrium diagrams. In Figs. 13–1 to 13–5 are shown a number of simple phase diagrams. Figure 13–1 is a

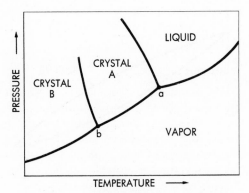

FIG. 13–1. Single-component equilibrium diagram.

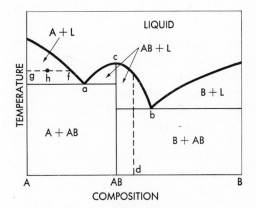

FIG. 13–2. Two-component equilibrium diagram.

single-component diagram with pressure as the ordinate and temperature as the abscissa. At point a, for example, there are three phases, crystal A, liquid, and vapor. Using the phase rule,

$$P + V = C + 2$$
$$3 + V = 1 + 2$$
$$V = 0.$$

Thus a is an invariant point with no degrees of freedom. Along the line ab, there are two components, crystal A and vapor, so here there would be one degree of freedom, either pressure or temperature. In any of the fields there would be only one phase, giving two degrees of freedom, both pressure and temperature.

In ceramic systems the vapor phase is not generally of importance. Thus it is convenient to work at a fixed pressure of one atmosphere and have what is known as the condensed system. This permits us to use the so-called isobaric phase rule, $P + V = C + 1$.

In Fig. 13–2 there is shown a two-component diagram with one compound AB and two eutectics. In the same way as for the previous diagram it may be shown that the eutectic points a and b have three phases, liquid and two solids, consequently

$$P + V = C + 1$$
$$3 + V = 2 + 1$$
$$V = 0.$$

If a composition AB cools, it will completely crystallize at point c and the solid phase will cool in this form. However, if a composition at d is cooled, the first crystals will come out at the liquidus curve, and the amount of crystals will increase as the temperature falls until the solidus line is reached. There the last of the liquid phase disappears.

The relative amounts of crystal and glass in a field may be determined graphically as follows. Take, for example, point h and draw a horizontal line intersecting the liquidus boundary. The ratio of solid to liquid is then hf/hg, by the so-called lever principle.

In Fig. 13–3 is shown a two-component diagram with the compound AB decomposing before it reaches the liquidus curve. This is called incongruent melting.

Figure 13–4 shows a two-component diagram in which A and B form a continuous series of solid solutions. If a composition AB_2 cools, crystals will begin to form at the liquidus curve, and these will have the composition AB_1. As the temperature drops

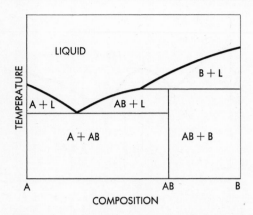

FIG. 13–3. Two-component equilibrium diagram with a compound *AB* melting incongruently.

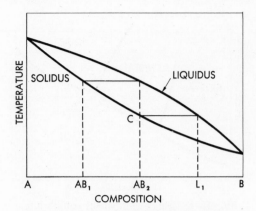

FIG. 13–4. Equilibrium diagram with a continuous solid solution.

the crystals will gain in *B* until at point *c* the last liquid disappears, and the crystal has the composition AB_2. The last drop of liquid will then have the composition L_1. Therefore, during the cooling there will be a constant interchange between the liquid and the crystals.

This discussion of the phase diagrams is very brief, only touching on some of the important points. The student is urged to read the excellent presentation of this subject by Hall and Insley, particularly with regard to three-component diagrams, a subject which cannot be covered here.

Example of an equilibrium diagram. As an example of the equilibrium diagram in the ceramic industry, the SiO_2-Al_2O_3 binary system is most appropriate. As shown in Fig. 13–5 there is one compound in the system, mullite, that melts incongruently at 1810°C. The eutectic lies close to the silica end. It would be expected from this diagram that a brick with 70 per cent alumina would contain some glassy phase on cooling down to the eutectic temperature of 1545°C, while a brick of 74 per cent alumina would have no glass content below 1810°C. Ac-

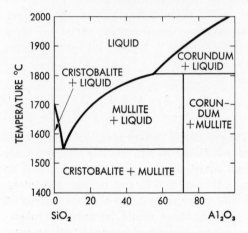

FIG. 13–5. The silica-alumina equilibrium diagram.

tually it is true that bricks a little below the mullite composition deform under loads above 1500°C, while those a little higher in alumina stand up well to 1750°C.

In making silica brick, it has been found that a more refractory product may be made by washing the gannister to remove a small portion of alumina. The reason for this is made clear by noting on the diagram that only one per cent of alumina drops the liquidus temperature 15°C.

Reaction rates

Chemical reactions in general proceed at a faster rate as the temperature is increased. This relation was expressed by Arrhenius as

$$\log \frac{k_{t_2}}{k_{t_1}} = A \left(\frac{1}{T_1} - \frac{1}{T_2} \right),$$

where k_{t_1} is the reaction velocity at temperature T_1, k_{t_2} is the reaction velocity at temperature T_2, and A is a constant (T is in the absolute temperature scale).

The reactions involved in firing a ceramic body follow this general law, so that a piece held at temperature for one hour would require a higher temperature than a piece held for 10 hours to arrive at the same degree of maturity. For example, Fig. 13–6 shows the porosity of a vitrified body fired for various lengths of time and at different temperatures. Each tenfold increase in time corresponds to a 23°C decrease in temperature. This corresponds to the constant A in the Arrhenius equation of about 9000, a value quite closely adhered to in many ceramic reactions.

Solid state reactions

Nearly all reactions in the traditional ceramic bodies were accompanied by some liquid that acted as a means of atomic transfer. More recently it has been believed possible to carry on reactions between two solids with no liquid phase. It seems now that such reactions may occur, but there is always the possibility that they may be sparked by a trace of impurity.

Single components. A solid state reaction may occur in one component because of a crystal growth. This is due to the loosening of atomic bonds by thermal energy so that the free atoms or atom groups tend to fall on the larger crystal faces having greater surface energy. Therefore, the small crystals grow smaller, and the large ones larger.

It should be kept in mind that not all reactions in ceramics follow the Arrhenius equation. For example, if MgO is formed by calcining $MgCO_3$ at various temperatures and for various times, the average size of the MgO crystal is shown by the curves of Fig. 13–7. While the size of the initial crystal formed is dependent on temperature, the growth rate of that crystal is the same for all temperatures and decreases with elapsed

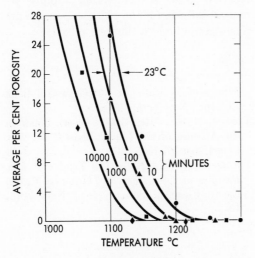

FIG. 13–6. Maturing conditions in a whiteware body.

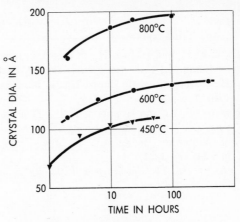

FIG. 13–7. Growth of MgO crystals from $MgCO_3$.

time. If the growth rates, not only for MgO but also for other oxides, taken from the slope of the previous curves are now plotted against time, a straight line on a log-log plot will result, as shown in Fig. 13–8. This relation may be expressed by

$$\log R = \log R_o - n \log t$$

$$R = R_o/t^n = \frac{31}{t^{1.56}},$$

where R_o is the rate of growth in Ångstroms per second when the time is one hour, and n is the slope of the straight line. This illustrates the usefulness of a log-log plot in determining the value of an exponential.

Two components. If two components, A and B, are present, a slow diffusion process may gradually form the crystal AB. The greater the surface, due to fine comminution, the more rapidly the reaction goes forward. Also if the bonds are loosened by the passing of one crystal through a transformation, the reaction rate is increased.

Some examples of solid reactions. A single-component reaction may be illustrated by the growth of MgO crystals as shown in the previous section.

A two-component reaction is the forma-

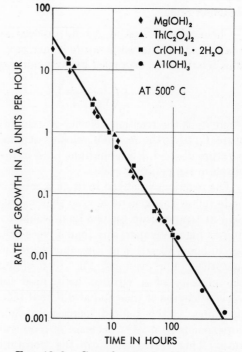

FIG. 13-8.   Growth rate of oxide crystals.

tion of spinel from MgO and Al_2O_3 at temperatures below the melting point. This reaction goes to completion rapidly by mutual diffusion if the two components are finely ground and intimately mixed.

REFERENCES

HALL, F. P., and INSLEY, H., "Phase Diagrams for Ceramists." *J. Am. Ceram. Soc.* **30**, Part II, November, 1947; Supp. No. 1, **32**, 153, 1949.

NORTON, F. H., and HODGDON, F. B., "The Influence of Time on the Maturing Temperature of Whiteware Bodies, I." *J. Am. Ceram. Soc.* **14**, 177, 1931.

QUILL, L. L., *The Chemistry and Metallurgy of Miscellaneous Materials.* McGraw-Hill Book Co., Inc., New York, 1950.

TAYLOR, N. W., "Reactions Between Solids in the Absence of the Liquid Phase." *J. Am. Ceram. Soc.* **17**, 155, 1934.

THOMPSON, M. DE KAY, *The Total and Free Energies of the Oxides of Thirty-two Metals.* Electro-Chemical Society, Inc., New York, 1942.

CHAPTER 14

TYPES OF CERAMIC BODIES

Introduction

Ceramic bodies vary widely in composition, depending on the finished properties required. A common brick might be made from many materials and still have sufficient strength, but the matter of cost requires that it be made of easily accessible clays that will fire at a low temperature. On the other hand, even though the cost is high, fine porcelain must be made of materials that are free from coloring impurities and from which will develop a body of high translucency.

The student should not attempt to memorize a series of batch formulae for bodies, but rather should understand the influence of the various ingredients on the properties. The present chapter is written with this idea in mind.

To give a quick view of the production of ware from various bodies in the United States, the chart in Fig. 14–1 has been prepared. It is impossible to make it complete but nevertheless it serves the purpose.

Methods of expressing the composition of a body

Batch formula. In the plant the body is expressed as the weights of the dry ingredients. A Sèvres porcelain body follows:

This method is convenient for compounding, but it does not permit easy comparison of one body with another.

Mineralogical formula. As a means of comparing bodies, it has been customary in Europe to express the body composition in equivalent minerals as first suggested by Segar. For example, the raw materials in the body are mixtures of minerals which may be computed from the chemical analysis. While this method is not used to a great extent in this country, due largely to doubts about the validity of the calculation of the "clay substance," it is of considerable value in obtaining a broad view of body compositions.

The Sèvres body given in the preceding section may be converted to the mineralogical formula if the following composition of the ingredients is known from the chemical analyses:

Zettlitz kaolin,
    95% clay substance, 5% quartz;
Quartz, 100% SiO_2;
Feldspar,
    88% KNa spar, 2% Ca spar, 10% quartz;
Marble, 100% $CaCO_3$.

Then a table is made up as follows:

Materials	Weights, kg	Per cent
Zettlitz kaolin	219.0	69.5
Quartz (potter's flint)	26.8	8.5
Feldspar (potash)	48.8	15.5
Marble ($CaCO_3$)	20.5	6.5
Total	315.1	100.0

Materials	Clay substance	Quartz	KNa spar	$CaCO_3$
69.5 Zettlitz kaolin	66.0	3.5	---	---
8.5 Quartz	---	8.5	---	---
15.5 Feldspar	---	1.6	13.6	0.3
6.5 Marble	---	---	---	6.5
Total	66.0	13.6	13.6	6.8

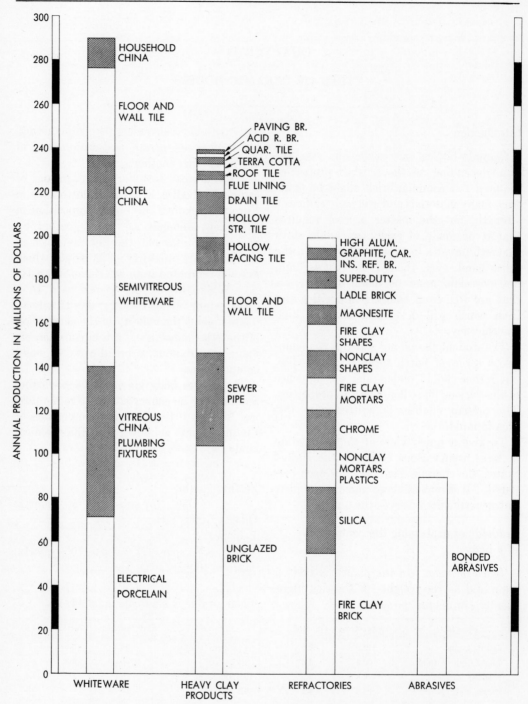

FIG. 14–1. Value of commercial ceramic bodies produced in the year 1947.

Equivalent formula. The preceding mineralogical formula may be converted as follows into an equivalent formula:

Materials	SiO_2	Al_2O_3	KNaO	CaO
Clay substance $\frac{66.0}{258} = 0.255$	0.510	0.255	---	---
Quartz $\frac{13.6}{60.1} = 0.226$	0.226	---	---	---
Feldspar $\frac{13.6}{54.7} = 0.025$	0.150	0.025	0.025	---
Calcium carbonate $\frac{6.8}{100} = 0.068$	---	---	---	0.068
Totals	0.886	0.280	0.025	0.068
Using Al_2O_3 = unity	3.16	1.00	0.089	0.24

The equivalent formula is then

0.089 KNaO

$\qquad$ 1.00 Al_2O_3 $\qquad$ 3.16 SiO_2.

0.24 CaO

Chemical composition. The fired or dried body may be analyzed by the usual procedure. In the case of the dried body given in the preceding sections the results are:

Silica	52.9
Alumina	28.9
Ferric oxide	0.5
Lime	4.0
Magnesia	0.2
Potash	1.7
Soda	0.7
Water (combined)	9.1
Carbon dioxide	2.5
Total	100.5

This includes the minor constituents Fe_2O_3 and MgO not shown before.

Triaxial whiteware bodies

These bodies are the usual ones for whitewares of various types. They are called triaxial because they are mainly clay, feldspar, and quartz, but all of them contain some alkaline earths, either as impurities or by design.

Semivitreous earthenware. This body is defined as having an absorption of 4 to 10 per cent and no translucency. It is usually white, but occasionally colored, and matures at cone 8 to 9 (1225 to 1250°C). A typical batch formula is shown in Table 14–1. This body is used for the bulk of the tableware made in the United States. It is durable and lends itself to rapid manufacture.

Hotel china. This is a body developed in this country to give great impact strength and still permit simple manufacturing operations. It has an absorption of 0 to 0.3 per cent and is translucent in thin sections. The maturing condition is represented by cones 9 to 11 (1250 to 1285°C). The body is generally white, but sometimes is colored by stains. A lighter weight ware, called household china, employs substantially the same body. The composition is shown in Table 14–1, where the added calcium carbonate (often dolomite) is distinctive.

Low-frequency electrical porcelain. This body is used for the bulk of the electrical porcelain made today. It matures at cone 12 (1310°C) with zero absorption, the glaze and body being fired together. The composition also is given in Table 14–1.

Hard porcelain. This body varies a good deal, depending on the temperature of firing.

Table 14-1

Composition of Whiteware Bodies
(per cent)

Type of body	Maturing at cone	Kaolin	Ball clay	Feldspar	Flint	Calcium sulphate	Barium carbonate	Frit	Bone ash	Talc	Kyanite	Fe$_2$O$_3$	MnO$_2$
Vitreous sanitary	12	28	20	32	20								
Electrical insulator	12	21	25	34	20								
Wall tile - vitreous	10	27	29	33	11								
Hotel china	10	34.8	7	22	35	1.2							
Semivitreous white-ware	9	24	28	13	35								
Bone china	10	13		15	32				40				
Parian	8	30	10	60									
Hard porcelain	16	35	15	25	25								
Belleek	8	35	15		20			30					
Basalt	10	30	15									36	19
Jasper	10		30		7		63						
Talc body	12	18		7						75			
Talc body	5	18	15	35	25					7			
Dental porcelain	10	5		81	14								
Refractory porcelain	20	50	10	10	5						25		

The more highly fluxed bodies mature as low as cone 10 (1260°C), while the magnificent Copenhagen porcelain is fired to cone 18 or 20 (1485–1520°C). The biscuit is fired to a low temperature for strength in handling and the glost fire is high, with a reducing condition at the end of the schedule to reduce the iron, thus producing a denser body and a whiter color. Several compositions are shown in Table 14–1.

Wall tile bodies. These are both porous and vitreous structures of the triaxial type. Recently talc has been much used in the porous ones. A typical vitreous body maturing at cone 10 (1260°C) is shown in Table 14–1.

Vitreous sanitary body. This is a one-fired body maturing at about cone 12 (1310°C) with a composition much like electrical porcelain.

Parian porcelain. This body, much used in small sculptured pieces, is named Parian because of the self-glazing property that gives it the appearance of marble (from the Greek island of Paros). It is composed of kaolin and feldspar, or sometimes frit, maturing at about cone 8 (1225°C) to zero absorption.

Dental porcelain. This body is largely feldspar with some flint and a few per cent of clay. Like Parian ware, it is self-glazing and nonporous.

Summary. In Table 14–1 there are listed a few typical whiteware bodies to give an idea of their relative composition. Remember, though, that many practicable bodies depart from these figures.

The diagram in Fig. 14–2 is plotted inside an equilateral triangle so that any point represents a total of 100 per cent of the three components, clay, quartz, and feldspar. The areas representing the various

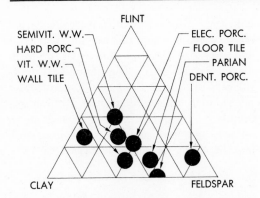

FIG. 14–2. The composition of various triaxial whiteware bodies.

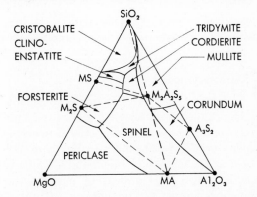

FIG. 14–3. Three-component diagram of Al_2O_3-MgO-SiO_2. The primary fields are solid lines and the compatibility triangles are dotted lines.

triaxial bodies are shown in their proper places. The composition of the body is designed to give the best properties with regard to plasticity, drying, firing, and the finished properties. The types of clay are varied to suit the molding process, whether it be pressing, jiggering, or casting.

Special electrical porcelain

Most of these bodies may be studied by the three-component equilibrium diagram of Al_2O_3-MgO-SiO_2 shown in Fig. 14–3.

Mullite. This body ($Al_6Si_2O_{13}$) has been used for spark plug cores because of its good mechanical strength and glass-free body. As this mineral melts incongruently, the mullite composition comes outside the field of primary crystallization. Mullite may be made from gibbsite or diaspore and kaolin, which are fused or calcined together. Or, kyanite may be used as the base, although this material does not contain enough alumina to produce 100 per cent mullite.

Magnesia or steatite. bodies. These fall in or around the field of clinoenstatite ($MgSiO_3$), as shown in Fig. 14–3. They are made by combining talc and clay in the proper proportions. Because they contain no alkalis, their electrical loss in a high-fre-

quency field is very low. The bodies have excellent strength but a rather short maturing range.

Cordierite bodies ($Mg_2Al_2Si_5O_{15}$). These come in the center of the field as shown in Fig. 14–3. They are notable for their very low coefficient of thermal expansion, but they have only a limited use.

Sintered alumina. The last valuable composition on the diagram is Al_2O_3, pure or mixed with a little clay. This body has very great strength and excellent heat shock resistance, and is a good high temperature insulator. However, very high temperatures are needed to mature it to zero absorption — cone 40 or 1850°C.

Zircon ($ZrSiO_4$). This compound mixed with some clay has been used in spark plugs for some time.

Titania porcelains. Within the last ten years the titania porcelains, especially the alkaline earth titanates, have assumed great importance. Titania or the titanates may be compounded with talc or kaolin to form a nonporous body. When freed of the alkalis, these bodies have the low electrical losses of the steatite porcelains, and in addition have dielectric constants as much as a thousand times greater. This high value permits the construction of very compact condensers.

Refractory bodies

Commercial bodies. Firebrick are usually compounded from plastic fire clay and flint clay or grog fired at 1250 to 1400°C to give a porosity of 15 to 20 per cent. Diaspore clay is used for super-duty brick, and alumina is added when it is desired to increase the alumina content still further.

Silica brick are made from gannister with 2 per cent of lime as a bond. The firing is carried out for a long period at 1500 to 1550°C to effect as much conversion as possible of the quartz to cristobalite and tridymite.

Magnesite brick are made from deadburned magnesite fired at 1600°C to give a porosity of 20 per cent. Chrome brick, made from chromite, are fired to a porosity of about 18 per cent. Both these types of brick are chemically bonded without firing for some uses.

Special oxide bodies. The most important oxide body is Al_2O_3, as mentioned previously. However, nonporous bodies of magnesia, stabilized zirconia, and beryllia are used in small quantities for melting crucibles and other laboratory ware. It should be kept in mind that beryllia is extremely toxic, and should be fabricated only under suitable conditions.

Other refractory bodies. There has been some interest in other bodies, such as carbides, nitrides, sulphides, silicides, and borides. Many of these compounds are highly refractory but not very resistant to oxidation. An exception is silicon carbide, which is used in considerable quantities in kiln furniture and refractory brick.

Bodies for thermal insulation. The production of thermal insulation is an important industry. Medium temperature insulators may be made from magnesium carbonate and asbestos, slag, or glass wool, or from diatomaceous earth. High temperature insulators are made from clay porosified by burned out organic matter or by fine gas bubbles. The porosity may be as high as 80 per cent and some of these insulators may be used to a temperature of 1650°C.

Special tableware and artware bodies

Bone china. This unique body is composed of nearly one-half bone ash (calcium phosphate) together with smaller amounts of kaolin, Cornwall stone and flint (Table 14–1). Because of the small amount of clay the body is low in plasticity and dry strength, and therefore requires a high degree of skill to fabricate. In firing, the deformation is so great that each piece must be set in a separate bed of flint. England is the only place where bone china has been made to any extent, and it is still produced as their fine china, in spite of the difficulties of manufacture. This is due partly to tradition and partly to the excellence of the fired body in strength, translucency, and color. The ware is matured at about cone 10 (1260°C) to an absorption of 0.3 to 2.0 per cent.

Frit porcelain. When Europe received the first pieces of Chinese porcelain, attempts were made to reproduce it. The first results were pieces made by mixing powdered glass with white earthenware body to produce a highly translucent, but not very strong body. Since that time frit porcelains have been improved and made in a number of places, such as Beleek, Ireland (Table 14–1). The fine china now made by Lennox in this country is a frit porcelain.

Jasper ware. This body, developed by Wedgwood in the middle of the 18th century for cameo ware, is composed of more than half barium sulphate, together with kaolin, ball clay and a little flint (Table 14-1). It is used mainly for biscuit ware, since it develops excellent colors with body stains.

Basalt ware. This jet black biscuit body is much in demand among collectors. It has a large proportion of iron and manganese oxides that give an excellent black color under the proper conditions of firing.

Abrasive bodies

Grinding wheels are made of SiC or Al_2O_3 grains bonded together. This bond may be ceramic, like porcelain or glass. Recently other bonds, notably organic resins, have been extensively used.

Stoneware bodies

Much of the stoneware made in the past was produced from natural stoneware clays containing some feldspar and fine silica. These clays burned to a dense, nonporous body. Most modern stoneware is compounded from clays, feldspar, and flint, which gives greatly improved properties. The distinction between a stoneware body and a porcelain body is mainly that the latter is largely free from iron oxide and therefore translucent. However, the stoneware may grade into porcelain.

Heavy clay product bodies

Bodies of natural clays and shales. Because of the low price of the product, most bricks and structural tile are made from a single clay, although in some cases sand may be mixed with the clay to reduce its shrinkage. These products are fired to a variety of temperatures depending on the impurities of the clays. The glacial brick clays of New England are fired to 1000°C, while some of the low grade fire clays used in the Middle West for brick are fired to 1250°C. Highly porous bodies do not withstand freezing well and low porosity bodies do not hold mortar, so the absorption is generally kept between 5 and 15 per cent.

These bodies vary greatly in color, depending both on the clay and the firing conditions. Many shades of red, buff, brown, and black are common.

Terra cotta bodies. These bodies are made from a plastic clay mixed with 15 to 35 per cent of coarse grog. They are fired at 1100 to 1300°C to produce a body with an absorption of about 15 per cent.

Properties of bodies

It seems appropriate to sum up this chapter with a list of the properties of the fired bodies so that a comparison between them may be made readily. However, there is space for only a few typical bodies, and not all their properties are accurately known.

Mechanical properties. In Table 14-2 are listed a few bodies for which strength is important, together with some of their mechanical properties. The superiority of sintered alumina is quite evident.

Thermal properties. In some cases the thermal properties of ceramic bodies are very important, especially in the case of refractories. In Table 14-2 are also shown a few thermal properties of some typical bodies.

Electrical properties. In Table 14-2 are given a few of the electrical properties of some insulator bodies. The unique property of the titania bodies is evident.

Chemical properties. The stability of bodies is sometimes of importance in the chemical industry, or under certain combustion conditions.

Optical properties. Color and translucency are important characteristics of fine porcelain. Unfortunately, there have been few quantitative measurements made on porcelains.

Table 14-2

Properties of Some Typical Bodies

Property at room temperature	Body					
	Stoneware	Steatite porcelain	Mullite porcelain	Zircon porcelain	Titania porcelain	Sintered alumina
Compression strength lb/sq in	75,000	100,000	100,000	120,000	---	400,000
Transverse strength lb/sq in	10,000	16,000	26,000	30,000	---	50,000
Elastic modulus, M, millions lb/sq in	12	14	15	25	---	52
Coefficient of exp. per °C	4×10^{-6}	9×10^{-6}	5×10^{-6}	5×10^{-6}	7×10^{-6}	6×10^{-6}
Thermal conductivity, K, cgs units	0.006	0.008	0.007	0.01	---	0.03
Dielectric constant, k	---	6	6	9	100	6
Loss factor	---	0.01	0.03	0.01	0.01	0.01

REFERENCES

BINNS, C. F., *Manual of Practical Potting*, 5th ed.   Scott, Greenwood and Son, London, 1922.

COLLINS, P. F., "Study of China Bodies of Belleek Type." *J. Am. Ceram. Soc.* **11,** 706, 1938.

LIN, C. C., "A Study of High-fire Porcelains." *J. Am. Ceram. Soc.* **2,** 622, 1919.

RIDDLE, F. H., "Ceramic Spark-Plug Insulators." *J. Am. Ceram. Soc.* **32,** 333, 1949.

SEARLE, A. B., *An Encyclopedia of the Ceramic Industries.*   Ernest Benn, Ltd., London, 1929.

SEGER, H. A., *Collected Writings.*   Chemical Publishing Co., Easton, Penna., 1902, pp. 669–708.

WATTS, A. S., "Some Whiteware Bodies Developed at the Ohio State University." *J. Am. Ceram. Soc.* **10,** 148, 1927.

CHAPTER 15

THERMOCHEMICAL CHANGES IN CLAYS AND BODIES

Introduction

The changes that take place in ceramic materials on heating are of vital importance to a thorough understanding of the firing operation.

Methods of measuring pyrochemical changes

When a ceramic material is heated there are many manifestations of change, the more important of which will be described below, using kaolin as an example.

Shrinkage. Most dry clays or bodies shrink on heating since the pores close up because of the surface tension forces pulling the particles together. Shrinkage or expansion may also occur if a mineral decomposes or inverts to another form. The shrinkage may be measured continuously by dial indicators or by reading telescopes, as a sample is slowly heated. The specimen may also be heated to successive temperature levels and measured when cooled between each cycle.

Figure 15–1 shows a linear shrinkage curve for a kaolin. The characteristics of the shrinkage may be made more clear if a rate of shrinkage curve is plotted from the

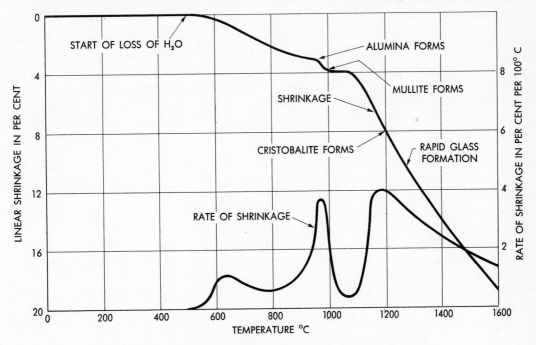

FIG. 15–1. Linear shrinkage and rate of shrinkage for a kaolin.

129

slope of the preceding curve. It will be seen that there are several regions of rapid shrinkage, which will be identified later.

Porosity changes. The porosity of a body is of importance, for it serves as an excellent measure of maturity. Porosity may be measured by the volume of water or air needed to fill the pores, or, if there are some closed pores, by grinding the sample so fine that all pores are broken open, and then calculating the solid volume from the weight and specific gravity.

These methods may be clarified by the following equations:

$$P = 100 \frac{V_p}{V_B} = 100 \frac{W_s - W_D}{W_s - W_u}, \quad (15\text{-}1)$$

where P is the per cent porosity, V_p the pore volume in cc, V_B the bulk volume in cc, W_s = sample weight saturated with water, W_D = sample weight dry, W_u = sample weight saturated and immersed;

or,

$$P = 100 \frac{V_p}{V_B} = 100 \left(1 - \frac{d}{s.g.}\right) \quad (15\text{-}2)$$

where $s.g.$ is the true specific gravity, and d is the bulk density.

Heat evolution and absorption. Again taking kaolin as an example, the heat release or absorption is shown in Fig. 1–9. The large heat absorption comes at about 600°C and a sharp evolution at 1000°C. The curve is obtained by heating the kaolin at a steady rate beside a neutral material and recording the temperature difference between the two. This is a useful method for analyzing clays, since each mineral writes its own signature.

Weight loss. If the same kaolin is heated slowly, or heated in steps with long periods of steady temperature, a curve of weight loss may be plotted as shown in Fig. 15–2. It will be seen that nearly the whole loss comes at a temperature of 450°C. The

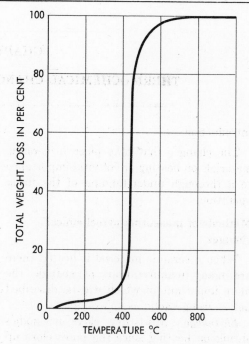

FIG. 15–2. Weight loss curve for a kaolin.

student should understand the reason for the difference between this temperature and the 600°C of the preceding paragraph.

Petrographic changes. Microscopic examination of materials after exposure to different temperature levels reveals much information when the crystals are large enough to be readily seen.

X-ray diffraction. In cases where the crystals are too small to be seen in the microscope, the x-ray technique gives the identity and often the size of the component crystals.

Summary. The results of the various methods, applied to kaolin, are tabulated in Table 15–1.

This accumulated information, then, allows us to picture quite clearly the changes taking place in kaolin when it is heated. The kaolinite crystals remain intact until 450°C is reached, and then they break down

Table 15-1

Thermal Changes in Kaolin

Temperature, °C	Shrinkage rate	Porosity	Heat effects	Weight loss	Petrographic results	X-ray diffraction results
20-100	zero	high	none	slight	kaolinite crystals	kaolinite crystals
100-400	zero	high	none	slight	kaolinite crystals	kaolinite crystals
400-500	slight	high	large absorption	very large	breakup of crystals	breakup of crystals
500-900	medium	high	none	slight	no visible crystals	amorphous meta-kaolin
900-1000	high	medium	large evolution	none	no visible crystals	γ-alumina and mullite appear
1000-1150	zero	medium	none	none	growth of mullite	growth of mullite
1150-1200	high	rapid decrease	small evolution	none	formation of cristobalite	mullite and cristobalite

substantially at constant temperature into an amorphous mass, although they retain the pseudomorph of the original crystals. At this point the H_2O is expelled with an absorption of heat and a loss of weight; however, the atomic structure does not seem to be greatly disturbed, since one oxygen atom is left from each two OH groups to tie the lattice together. This is confirmed by the reversible nature of the structure, as water can be put back to give kaolinite again.

At 980°C there is a sharp evolution of heat caused by sudden crystallization of the amorphous mass into mullite, although γ-alumina has also been found as a transitory phase. As the temperature increases, the mullite crystals grow, the glassy phase pulls the particles together, causing shrinkage, and at around 1200°C cristobalite crystallizes from the siliceous glass.

Effect of heat on clays and other hydrated materials

Kaolin. This clay has been discussed in the previous section.

Other clays. In general, all clays show the same effects on heating as kaolinite, al-

though they are influenced by impurities which dilute the lower temperature effects and reduce the temperature of glass formation. Some clays, such as flint fire clays, containing sulphates tend to bloat as soon as a glass phase is formed and therefore show little shrinkage, and at times even an expansion.

Clay has many impurities that break down individually on heating. In Table 15-2 a few of them are shown.

Gibbsite. The thermal curve for gibbsite is shown in Fig. 1-9. The OH groups are

Table 15-2. Decomposition of Clay Impurities

Reaction	Breakdown temp. °C
$FeS_2 + O_2 \rightarrow FeS + SO_2$	350-450
$4FeS + 7O_2 \rightarrow 2 Fe_2O_3 + 4SO_2$	500-800
$Fe_2(SO_4)_2 \rightarrow Fe_2O_3 + 3SO_3$	560-775
$C + O_2 \rightarrow CO_2$	350→
$S + O_2 \rightarrow SO_2$	250-920
$CaCO_3 \rightarrow CaO + CO_2$	600-1050
$MgCO_3 \rightarrow MgO + CO_2$	400-900
$FeCO_3 + 3O_2 \rightarrow 2Fe_2O_3 + 4CO_2$	800→
$CaSO_4 \rightarrow CaO + SO_3$	1250-1300

driven off at a comparatively low temperature because of the weak bonds. There is no exothermic peak at 980° since no silica is present to form mullite; thus γ-alumina is formed gradually and changes to α-alumina (corundum) at high temperatures.

Diaspore. The thermal curve of this mineral is also shown in Fig. 1–9. The crystals are broken up at a higher temperature than was the case for gibbsite, but the final crystals are the same, first γ-alumina, then α-alumina.

Montmorillonite. This mineral, described in Chapter 1, breaks down at 600°C to 800°C with the final formation of mullite and glass.

Hydrous mica (illite). This mineral behaves in much the same way as montmorillonite.

Thermal changes in crystals with inversions

Many crystals may occur in several allotropic forms which invert from one to another with change of temperature. Two types of inversion are recognized; one is the reversible type where a very small atomic rearrangement is needed, and the other is the irreversible, sluggish type which requires a considerable rearrangement of the atoms. As these inversions are often accompanied by considerable volume changes, they are of great interest in ceramics.

Silica. Silica in its many forms has been studied a great deal by both the geophysicist and the ceramist. There are at least six crystalline forms of silica, all composed of a three-dimensional network of silicon-oxygen tetrahedrons sharing corners, so that every oxygen atom is shared by one silicon atom to give the formula SiO_2.

The densest form of silica is quartz, which inverts reversibly at 573°C to the high form. A more loosely packed structure is tridymite, with inversions at 117°C and 163°C. A still less dense type of packing gives cristobalite with an inversion at about 250°C. In all these reversible inversions there is a slight rotation of the tetrahedrons but no breaking of the Si-O linkages.

On the other hand, when quartz is maintained above 870°C for a long time, it slowly transforms to tridymite which in turn changes to cristobalite above 1470°C. These changes require a major rearrangement of the structure, with some bonds breaking and others forming. This is a relatively slow process and one that is irreversible. Impurity atoms that tend to distort the silica lattice greatly accelerate the trans-

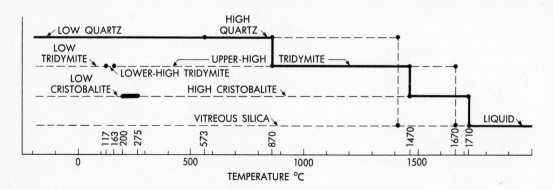

FIG. 15–3. The forms of silica with their stability regions.

formation. For example, lime accelerates the inversion of quartz to tridymite and cristobalite in firing silica brick.

The silica system is summed up in the graphic diagram of Fig. 15–3, taken from Sosman.

Kyanite. Kyanite when heated to 1100°C starts to decompose into glass and mullite, quite irreversibly. This decomposition is accompanied by a volume increase of considerable magnitude.

Pyrochemical changes in triaxial bodies

Life history of a whiteware body. When a mixture of the three components, clay, feldspar, and quartz, is heated slowly, the following changes are found to take place. Below 950°C the only changes are the inversion of low to high quartz and the breakup of the clay minerals, with a little shrinkage. At 1000°C, mullite has started to form in the clay grains, and glass formation starts to pull the mass together. Also, the quartz starts to invert to cristobalite. At 1100°C still more glass is formed by solution at the edges of the feldspar grains, and the pores decrease in number and assume a spherical shape. At 1150°C the pores are still smaller; some solution of the silica grains occurs with more glass phase. At 1200°C the body is mature with zero porosity. Inside the pseudomorphs of the feldspar grains may be seen large mullite needles formed by the diffusion of alumina from the clay through the glass.

Translucency. This property is a very desirable one for fine tableware. Translucency is increased by increasing the glass phase to any desired degree, but practically a high glass phase means deformation in firing and poor physical properties. It is obvious that if the glass phase has the same index of refraction as the crystals, the translucency is increased since there is less scattering of transmitted light. In the whiteware body there are chiefly quartz and mullite crystals. As the former has an index of refraction of 1.54, while for the latter the average value is 1.64, it is obviously impossible to get a glass to match both. However, if either one is matched a real increase in translucency results as shown in Fig. 15–4. Unfortunately, it is difficult to use glass with a high index of refraction since it is so reactive that it dissolves silica and thus lowers its own index. Much more work could be done on this interesting problem.

Thermal reactions in refractory bodies

Grog and clay bodies. In this case the grog is generally stabilized at temperatures at least as high as those at which the body is fired, so no reaction occurs in this mate-

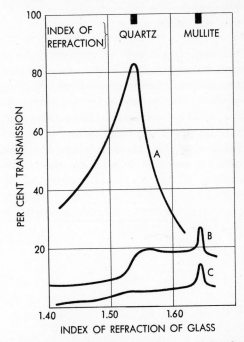

FIG. 15–4. Showing the influence of the glass phase on translucency: (a) 40% quartz, 60% glass; (b) 20% quartz, 20% mullite, 60% glass; (c) 30% mullite, 70% glass.

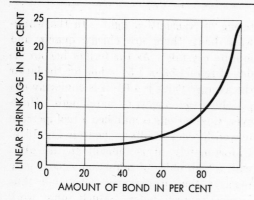

FIG. 15–5. Shrinkage values for grog-clay mixtures.

rial. The reactions in the clay are similar to those described for kaolin. The grog, however, has a mechanical effect in that it replaces a volume of the clay lost by shrinkage. Thus the total shrinkage of the body is reduced. The curve shown in Fig. 15–5 is most interesting. This represents the linear shrinkage of a complete series of mixtures of grog and clay. The 100 per cent grog mixture shows some shrinkage as the sharp contact points frit together. As clay is added, the shrinkage remains constant up to a 30 per cent content of clay, at which point the pores between the grog particles are just filled with clay. The addition of any more clay separates the grog particles further and further; this results in ever increasing shrinkage until 100 per cent clay is reached.

Glass-free refractory bodies. There has been great interest recently in the means by which pure oxide bodies sinter to a non-porous condition. This process seems to be a combination of surface and bulk flow which gradually closes up the pores and forms a dense mass. Lattice defects caused by initial straining of the crystals seem to accelerate this process.

Metal-ceramic bodies. Since the Second World War, there has been much interest in bodies composed of mixtures of metals and ceramics, such as alumina with chromium. It was hoped that these mixtures would combine the good properties of metals and ceramics, but as yet no outstanding results have been made public.

Some special bodies

Bone china. This body, on heating, forms a large proportion of calcium phosphate glass; this surrounds the flint, the Cornwall stone particles, and the mullite from the clay. The high refractive index of the phosphate glass and the relatively small amount of mullite account for the high degree of translucency of this body.

Frit porcelains. In these bodies the frit melts and dissolves the feldspar and part of the clay and quartz. The large amount of glass causes the body to be rather soft at the maturing temperature and to require good support in firing.

REFERENCES

COMEFORO, J. E., et al., "Mullitization of Kaolinite." *J. Am. Ceram. Soc.* **31**, 254, 1948.

GOODMAN, G., "Relation of Microstructure to Translucency of Porcelain Bodies." *J. Am. Ceram. Soc.* **33**, 66, 1950.

GREIG, J. W., "Formation of Mullite from Cya-nite, Andalusite, and Sillimanite." *J. Am. Ceram. Soc.* **8**, 465, 1925.

NORTON, C. L., "The Influence of Time on the Maturing Temperature of Whiteware Bodies, II." *J. Am. Ceram. Soc.* **14**, 192, 1931.

NORTON, F. H., "Critical Study of the Differential Thermal Method for the Identification of the Clay Minerals." *J. Am. Ceram. Soc.* **22,** 54, 1939.

NUTTING, P. G., *Some Standard Thermal Dehydration Curves of Minerals.* U. S. Geol. Survey, Prof. Paper 197–E, 1943.

SHELTON, G. R., and MEYER, W. W., "The Nature of the Glass Phase in Heated Clay Materials, II." *J. Am. Ceram. Soc.* **21,** 371, 1938.

SOSMAN, R. B., *Properties of Silica.* Reinhold Publishing Corp., New York, 1927.

CHAPTER 16

KILNS AND SETTINGS

Introduction

It is not expected that the ceramic student will soon become a designer of kilns, as this takes long experience in the field. However, it is necessary for him to know the principles underlying the construction and operation of the kiln, and these will be briefly covered in this chapter.

Principles underlying kiln design

The essential components. The kiln must have four main components: first, a base on which to set the ware; second, a source of heat; third, a means of transferring the heat from the source to the ware; and fourth, an envelope to confine the heat to the ware.

Periodic kilns. The earliest type of kiln was the so-called periodic kiln in which the firing process consisted of setting the kiln, firing up to the maturing temperature, cooling, and then drawing the ware. This cycle varied from a few hours in length for a small laboratory kiln to as much as a month for a large production kiln.

The earliest types of periodic kilns had the fire box below the ware, so that the hot gases could pass upward by natural draft. This is called the updraft kiln. However, it was found that the temperature uniformity was poor, not only because the bottom of the kiln was hotter than the top, but also because hot areas in the horizontal cross section tended to become still hotter due to the chimney action.

This lack of temperature uniformity was corrected to some extent by using a down-draft kiln in which the hot gases passed down through the charge and then up a stack. Kilns with horizontal draft were also used, as well as kilns with flues in the walls to give a combined up-and-down draft principle. In cases where the charge was glazed, as with terra cotta, muffle kilns were developed in which the heat from the combustion gases passed to the ware by conduction through thin but gas-tight walls. Figure 16–1 shows diagrammatically a number of types of periodic kilns.

The thermal efficiency of periodic kilns depends on many factors, such as maximum temperature, temperature uniformity, type of setting, insulation, and heat capacity of the kiln itself. The diagram in Fig. 16–2 shows the heat balance for a typical periodic kiln for firing refractories. A large part of the heat provided by the fuel is lost in the flue gases and in the cooling bricks, allowing only 20 per cent for actually firing the ware.

Continuous chamber kilns. In order to save fuel, many efforts were made to develop kilns that used a portion of the great amount of heat wasted in the periodic kiln. In China, as early as the first century, chamber kilns were built up a hillside, so that the lower chambers exhausted into the ones above. In Europe during the middle of the 19th century the continuous chamber kiln was developed, effecting a substantial saving in fuel. These kilns were a series of periodic kilns, joined side by side, and fired in succession. This permitted drawing the exhaust gases from the kiln being fired

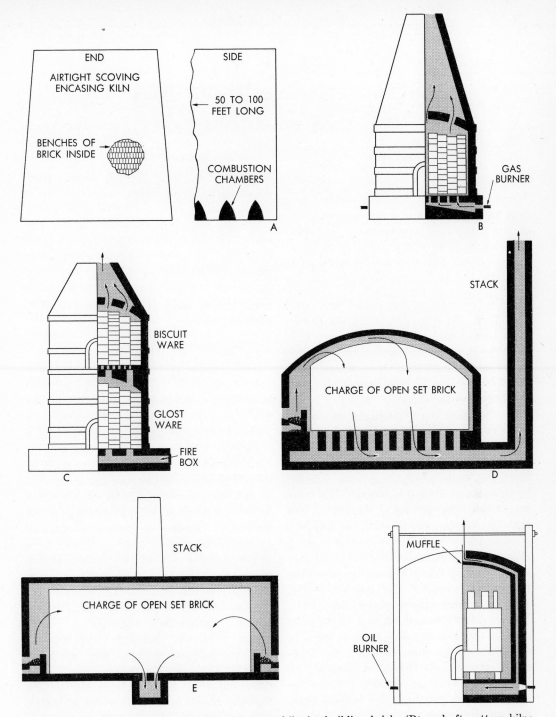

FIG. 16-1. Types of periodic kilns: (A) scove kiln for building brick; (B) updraft pottery kiln; (C) round updraft kiln for hard porcelain; (D) round downdraft kiln for fire brick; (E) Newcastle type of horizontal draft kiln; (F) muffle kiln for terra cotta.

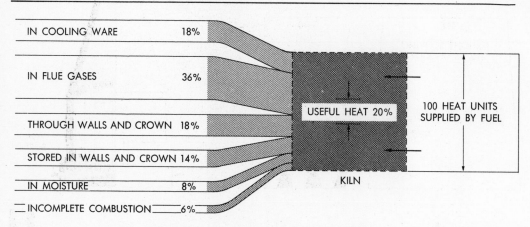

FIG. 16–2. Heat balance in a periodic kiln.

through the adjacent kiln, which was to be fired next, and thus preheating the charge close to the maturing temperature. Further, the gases, after passing through this kiln, could be used to preheat several more kilns or chambers, as they are called, ahead of the fire.

At the same time the chambers on the other side of the fired chamber were cooling, and by pulling air through them, preheat could be provided for combustion. Thus, the firing would move along from chamber to chamber in a regular manner. In order to prevent interruptions in the cycle when the end chamber was reached, the kiln was often arranged in the form of a ring; in some cases so many chambers were provided that several firing points could move around the ring simultaneously. The diagram in Fig. 16–3 should make the operation clear. The firing is usually carried out in a very simple manner by dropping coal through holes in the crown of the chamber under fire. Because of the high degree of preheat, no grates or fireboxes are needed. There are many variations in the design of these kilns, but the principle is the same in all of them.

The thermal efficiency of these kilns is high, but in using the expression for kiln efficiency of

Per cent efficiency = 100 ×

$$\left(\frac{\text{Heat units required to heat the charge}}{\text{Heat units input in fuel}}\right),$$

it is found that values of over 100 per cent often occur. This simply indicates that such a formula is not suited for use with a recuperative system. As it is often difficult for the student to comprehend the meaning of recuperation in the calculation of efficiency, the simple diagrams of Fig. 16–4 have been prepared to indicate the effect of increasing amounts of recuperative action on the calculated efficiency.

Tunnel kilns. It will be obvious that a recuperative kiln may be designed similar in principle to the chamber kiln except that the fire remains in one place and the ware moves continuously through it. This change has two important advantages: first, the kiln structure at any one point stays at the same temperature for a long time and thus is freed from the strains brought about by a periodically changing temperature, as is the case in the chamber kiln; and second, the ware is mounted on

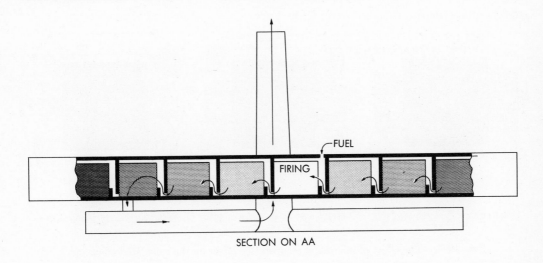

SECTION ON AA

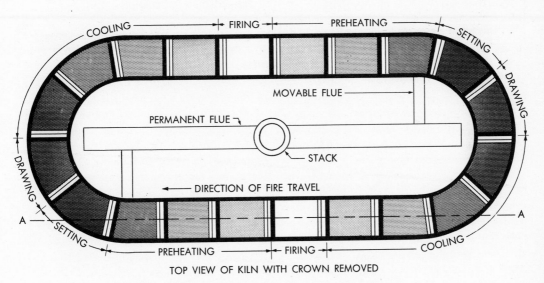

TOP VIEW OF KILN WITH CROWN REMOVED

FIG. 16–3. The firing operation in a chamber kiln.

cars and may be set and drawn in the open at a convenient place.

As the name implies, the tunnel kiln consists of a long refractory tunnel with a heat source in the center and a series of cars carrying the ware moving continuously through it. In Fig. 16–5 is shown a side view of such a kiln, together with curves of temperature and pressure throughout its length.

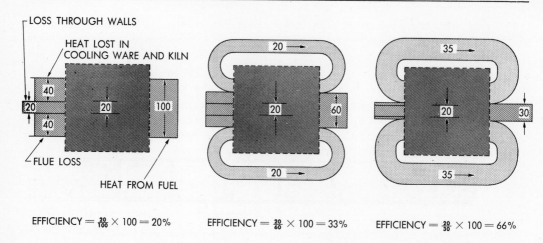

EFFICIENCY = $\frac{20}{100} \times 100 = 20\%$ EFFICIENCY = $\frac{20}{60} \times 100 = 33\%$ EFFICIENCY = $\frac{20}{30} \times 100 = 66\%$

Fig. 16–4. The effect of increasing recuperative effect on the calculated efficiency.

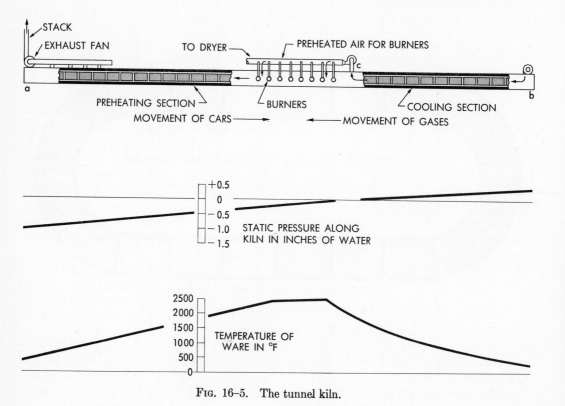

Fig. 16–5. The tunnel kiln.

The cars with the green ware pass into the kiln at *a*, usually through an airlock. Here they meet the low temperature combustion gases which have given up much of their heat to previous cars. As the cars reach the hot zone, their temperatures reach a maximum which may be maintained for some length of time. Beyond the hot zone they are slowly cooled by air entering the kiln at *b*. This air, gradually picking up heat, is drawn out of the kiln at *c* and a portion put back as preheated air for the burners.

A cross section of a tunnel kiln is shown in Fig. 16–6. It will be seen that the charge fits closely to the tunnel section so that the longitudinally flowing gases largely pass through the setting itself and thus effect an efficient heat interchange. The car top is made of refractory material thick enough to protect the metal car frame and wheels. Along each side of the inner wall of the kiln is a trough filled with sand. Into this dips a steel blade attached to the sides of the car, thus forming a gas-tight seal.

The thermal efficiency of the tunnel kiln is about the same as for chamber kilns, but there is an additional saving in the labor required for setting and drawing. A heat balance diagram for a tunnel kiln is approximated by the third diagram in Fig. 16–4. A large part of the losses are through the great area of the crown and walls, indicating the need for thorough insulation. Increasing the length of the kiln will decrease the exhaust gas temperature and therefore the heat loss, but a compromise must be reached by the designer.

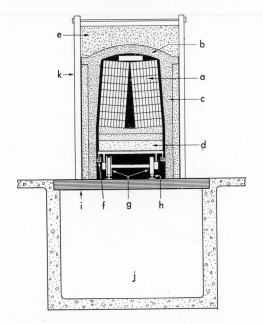

FIG. 16–6. Cross section of a tunnel kiln: (a) brick setting; (b) refractory crown; (c) refractory walls; (d) car top; (e) insulation; (f) sand seal; (g) wheels; (h) tracks; (i) supporting cross beam; (j) inspection tunnel; (k) buck stays.

Fuels and burners

There is not space in this book to consider the combustion process, but the student of ceramics should derive an understanding of this subject from other courses. However, a little will be said about fuels and their combustion in ceramic kilns and glass tanks.

Fuels. The early kilns were fired with wood — an excellent fuel, but one requiring much labor. Until quite recently the fine porcelain of Copenhagen was fired at 1500°C with dry wood.

Coal was widely used in the potteries, but now is largely confined to kilns for refractories and heavy clay products. A coal low in sulphur content and with a long flame is desirable.

Fuel oil is used in many kilns, especially tunnel kilns where the high thermal efficiency compensates for the high cost of the fuel. It is also used in many glass tanks.

Natural gas is much used in the pottery industry as it gives a clean, easily controlled heat. We are fortunate in this country still to have plentiful supplies of this fuel in many areas.

Producer gas is used to a small extent in this country for a fuel and in Europe is quite commonly used.

Burners. Coal is burned on grates except in chamber kilns, where it is dropped on the floor. Mechanical stokers are being used to an increasing extent to minimize the amount of labor.

Oil is atomized either by air or by mechanical means into a fine mist which is intimately mixed with the combustion air.

Gas is burned in a variety of burners, depending on the composition and the type of flame desired. Intimate mixing gives a short, intense flame, while little mixing produces long, luminous flames.

Fuel consumption

In Fig. 16–7 are given the heat units per pound of ware required to fire various types of ware in several different kilns. It will be noticed that tunnel and chamber kilns for bricks use about the same amount of fuel for a given temperature, and that periodic kilns use about twice as much. The use of muffles

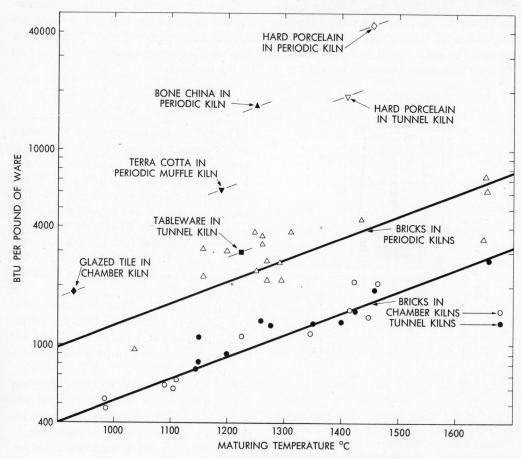

Fig. 16–7. Heat units required for firing in the ceramic industry.

nearly doubles the fuel consumption again, and setters or saggers greatly increase the fuel required per pound of ware. This method of expressing fuel consumption is the only one by which satisfactory comparisons can be made.

Examples of kilns

Periodic kiln. Figure 16–1(D) shows a typical round downdraft kiln used in the refractories and heavy clay products industry. The fuel is coal, burned in a series of fire-

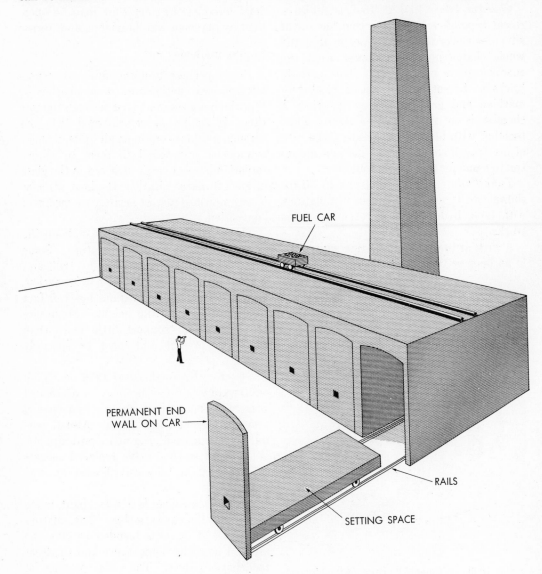

FUEL CAR

PERMANENT END
WALL ON CAR

RAILS

SETTING SPACE

FIG. 16–8. A chamber kiln with moving car bottoms.

boxes. The temperature uniformity is not very good in this type of kiln, but this does not greatly matter in the case of structural clay products and some types of refractories. The kiln is very wasteful of fuel, however, and can only be used where coal is cheap.

Chamber kiln. In Fig. 16–8 is shown a recent type of chamber kiln that has a real advantage over the older types in that the whole charge may be removed from the chamber on a car bottom. This permits bricks to be set and drawn in blocks by machine and greatly reduces the time a chamber is out of the heating cycle. This, together with blowers to speed up the rate of fire travel, should make this kiln an attractive one for use in this country.

Tunnel kiln. In Figs. 16–9 and 16–10 are shown two typical tunnel kiln installations with fans, hydraulic pusher, and transfer trucks.

A circular tunnel kiln for firing whiteware in an open setting is often used. Here the

FIG. 16–9. A tunnel kiln car showing an open setting of electrical porcelain (R. C. Remmey Son Company).

cars are in the form of a ring that revolves like a slowly moving merry-go-round. The small cross section assures an even temperature distribution and ease of setting and drawing from one location. These kilns are popular in the whiteware industry for small production, since they may be operated by one man who fires, sets, and draws.

Setting methods

Bricks. These uniform units are generally set in a checkerwork or in benches so that the hot gases may pass through the setting. It should be remembered that the ceramic products are in small units, so heat cannot be transferred to them by direct radiation in most cases, as it can in the glass or steel furnace. Rather, the heat transfer is accomplished almost entirely by means of convection.

The height of the setting depends on the type of brick and the temperature. Some brick, such as silica refractories, will support a high setting with no deformation. Others deform under medium loads, a fact that limits the setting height. Magnesite refractories will support little more than their own weight and must be specially boxed.

Saggers. Up to the year 1920 nearly all whiteware was set in fire clay boxes, called saggers. This protected them from uneven temperatures and kiln gases. Also it permitted setting each piece on its own support. While open settings have replaced saggers to a large extent, they are still used for some products.

The modern sagger is thin and light, made by casting in plaster molds. The mixture of clays and grogs is blended to give the greatest heat shock resistance and the least tendency to dust. The older practice of setting green saggers in the production kiln has largely been abandoned and the sagger

FIG. 16–10. General view of a tunnel kiln, showing blowers and ducts (Allied Engineering Company).

is prefired to a temperature higher than that encountered in use. The sagger today often has a life of over 100 trips through the kiln, as against 20 or 30 in the past.

In the biscuit firing the flat ware is generally stacked 12 to 15 high with a refractory setting ring at the bottom and sometimes at the top (Fig. 16–11(a). In the case of hotel china a filler is put between the plates to prevent sagging, as shown in Fig. 16–11(b). Hollow ware is boxed as in Fig. 16–11(c).

Bone china and frit porcelain must be placed on a flint bed, so that each piece requires a separate sagger. This bed is formed with a plaster tool pressed into the dry flint. On firing, the piece settles down to conform with the bed as shown in Fig. 16–11(d).

Hard porcelain is high fired with the glaze, so each piece must have its own sagger, resting on its unglazed foot as shown in Fig. 16–11(e).

Open settings. Open settings have several advantages over saggers. In the first place 50 to 100 per cent more ware may be set in a given kiln volume. Second, the weight of the open setting is somewhat less than that of the corresponding saggers, with a consequent decrease in fuel consumption. Third, in some cases the placing labor is reduced. On the other hand, open settings do not protect the glost ware from the combustion gases, and thus require muffle kilns or relatively flameless combustion. Also, there is more of a tendency for the glaze to volatilize, or become dry, in the open setting. In some cases a change of glaze composition is necessary to prevent this.

The average open setting consists of a series of slabs separated by corner posts. The ware is stacked on these slabs in various ways to obtain a close packing. In Fig. 16–11(f) is shown an open setting for cups.

The slabs and posts are made of various refractory mixes having stability, good thermal shock resistance, and good hot strength. A mixture of clay and silicon carbide is widely used at present.

Glost setting of tiles is shown in Fig. 16–11(g) and of semivitreous plates in Fig. 16–11(h). In Europe glost earthenware plates are reared up in saggers as in Fig. 16–11(i).

Special setting techniques. The setting of porcelain pieces, such as figurines, requires

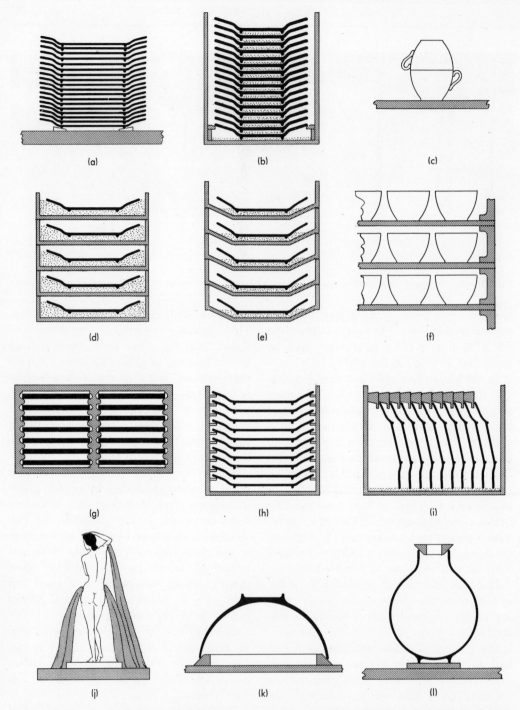

Fig. 16–11. Various setting methods: (a) semivitreous biscuit plates; (b) hotel china biscuit plates; (c) biscuit cups; (d) bone china; (e) hard porcelain; (f) glost cups in open setting; (g) glost tiles; (h) glost semivitreous plates; (i) glost plates; (j) figurine of Parian; (k) hard porcelain bowl; (l) jasper vase.

a skill that is very seldom available in this country. Figure 16–11(j) shows a porcelain piece supported with unfired props of the same body.

Cups and bowls must be fired on green setters if a perfect piece is desired. Several types of setters are shown in Figs. 16–11(k) and 16–11(l).

REFERENCES

BRONGNIART, A., *Arts Ceramiques*, Deuxieme Ed. Béchet Jeune, Paris, 1854.

DRESSLER, P., "Problems of Firing Ceramic Ware in Tunnel Kilns." *Bull. Am. Ceram. Soc.* **68**, 411, 1939.

GOULD, R. E., *The Making of True Porcelain Dinnerware*. Industrial Publications, Inc., Chicago, 1947.

HASLAM, R. T., and RUSSELL, R. P., *Fuels and Their Combustion*. McGraw-Hill Book Co., Inc., New York, 1926.

MANSUR, H. H., "The Heat Balances of Some Ceramic Kilns." *J. Am. Ceram. Soc.* **14**, 89, 1931.

NORTON, F. H., *Refractories*, 3rd ed. McGraw-Hill Book Co., Inc., New York, 1949, Chaps. 9 and 10.

ROSENTHAL, E., *Pottery and Ceramics*. Penguin Books, Harmondsworth, Middlesex, England, 1949.

TRINKS, W., *Industrial Furnaces*. John Wiley and Sons, Inc., New York, 1942, Vol. I (3rd ed.), Vol. II (2nd ed.).

CHAPTER 17

GLASSY STATE

Introduction

In Chapter 1 it was stated that solid matter exists either in the crystalline state, where the atoms are in an orderly array, or in the glassy state, where the atoms form a random three-dimensional network. It is the purpose of this chapter to examine the latter state of matter with regard to its nature, its stability, and its properties. It should always be kept in mind that the important characteristic of glass is its transparency.

Constitution of glass

Glass has been defined in simple terms as an undercooled liquid of very high viscosity, but a more precise definition is that given by Morey: "A glass is an inorganic substance in a condition which is continuous with and analogous to the liquid state of that substance, but which as the result of having been cooled from a fused condition has attained so high a degree of viscosity as to be for all practical purposes rigid." For many years the nature of glass was not understood. Only when x-ray diffraction techniques were brought to bear on this problem by Warren and his co-workers was it possible to obtain a clear picture, and then only for the simple glasses.

Glass network. The x-ray spectrum of a crystal has many sharp lines, but glass produces only a few very diffuse bands, showing that little order is present. The first diffuse band for glass represents the Si-O distance, the second, the O-O distance; after this the randomness of the structure precludes many other bands. It should be kept in mind that all the Si-O distances are not exactly alike because of the varying angles of the bonds and nonrepeating neighbors. This accounts for the great width of the bands.

The rather involved method of translating x-ray diagrams into the structure cannot be touched on here, but the student who is interested may find this information in the references. It is sufficient to say that the final conclusion may be illustrated by the rather conventionalized two-dimensional drawings in Fig. 1–1(b). The crystal has a uniform network with a unit repeated time after time in all directions, while the glass structure is a random network with only the silicon-oxygen tetrahedron a nearly invariable unit. In this random network there are holes of various sizes in which may rest other atoms such as sodium and calcium. Because of the randomness it is not necessary to have stoichiometric proportions of these additional atoms as it is in a simple crystal; on the contrary, they may occur over a considerable range as the holes are gradually filled.

On the other hand, there is no certainty that the random structure is uniform throughout; it is possible in a sodium silicate glass, for example, that small volumes might have the composition SiO_2 and other volumes the composition Na_2SiO_3 in such proportions as to give the complete glass composition. If randomness were maintained, the x-ray diffraction technique would not be able to show this segregation.

Much may be learned about the coordination number of cations in the glassy structure from its color, as discussed in Chapter 21.

Network formers. It has been known for a long time that certain elements and compounds could be produced in the glassy state, but it was not until the classic paper of Zachariasen appeared in the year 1932 that the mechanism of glass formation was made clear. Dr. Zachariasen stated four rules that should be fulfilled before an oxide can be considered a glass former. These are: (1) each oxygen atom must not be linked to more than two cations; (2) the number of oxygen atoms around any one cation must be small; (3) the oxygen polyhedra must share corners, not edges, to form a three-dimensional network; (4) at least three corners of each must be shared.

On the basis of these rules, the following oxides should in themselves form glasses: SiO_2, BeO_3, GeO_2, P_2O_5, and As_2O_5: Actually, all of these oxides will form glasses. And in addition, As_2O_3, Sb_2O_3, Bi_2O_3, and beryllium fluoride will form glass by the same criteria.

There are other materials with long chain molecules that form glasses, such as the elements sulphur, selenium, and tellurium, and the compounds meta-phosphoric acid, zinc chloride, and some of the sulphides. None of these, however, have great practical value.

A number of workers in the glass field have classified the cations into three classes; the network formers that can form a glass by themselves, the modifiers that cannot form glasses but can enter into the holes in the network, and the intermediates that at times may enter into the glass network to a limited extent. It has also been shown that the bond strength between a cation and an oxygen atom is an indication of glass-forming ability; the stronger bonds are the glass

formers and the weaker ones the modifiers. Table 17–1, as worked out by Sun, shows the single bond strength of some cations as related to their glass-forming ability. It will be seen that the cations arranged in groups according to bond strength give very good agreement with their glass-forming properties.

Network modifiers. The modifiers consist of the alkali ions, the alkaline earth ions, lead ions, zinc ions, and many others of less importance. As mentioned before, they fit into the holes of the network and loosen the bonding so that the glass has a lower softening point, is less chemically resistant, and has a greater coefficient of expansion.

Intermediate glass formers. There are other cations that may under suitable conditions enter the glass network itself as par-

Table 17-1

Calculated Bond Strengths of Oxide Constituents in Glasses

	Cation	Valence	Coordination no.	Single bond strength
Glass network formers	B	3	3	119
	Si	4	4	106
	Ge	4	4	108
	Al	3	4	101–79
	Be	3	4	89
	P	5	9	111–88
	V	5	4	112–90
	As	5	4	87–70
	Sb	5	4	85–68
	Zr	4	6	81
Intermediates	Zn	2	"2"	72
	Pb	2	"2"	773
	Al	3	6	53–67
	Zr	4	8	61
	Cd	2	"2"	60
Modifiers	Na	1	6	20
	K	1	9	13
	Ca	2	8	32
	Mg	2	6	37
	Zn	2	4	36
	Pb	4	6	39

tial substitutes for the network former. In other cases, they will actually form the glass network with other nonglass-forming substances.

In the first case some cations, such as Al^{+++}, may partially substitute for Si^{++++} in the network. This is what happens in the clay minerals. In the second case, we have examples of such glasses as those found in the $MgO-CaO-Al_2O_3$, the $K_2O-CaO-Al_2O_3$, and the $BeO-Al_2O_3$ systems. None of the cations in these systems are glass formers in themselves. Al^{+++}, which in Table 17–1 is just on the borderline of the glass former group, acts in these particular cases as a glass network former.

Table 17–2

Methods of Expressing Glass Composition

Purpose	Expression	
Mol per cent for glass	SiO_2	75.0%
	Na_2O	12.5%
	CaO	12.5%
Weight analysis	SiO_2	75.3%
	Na_2O	13.0%
	CaO	11.7%
Batch to yield 100 pts. of glass	Sand	75.4
	Soda ash	22.2
	Limestone	20.9
		118.5
Enamel frit	Sand	63.6%
	Soda ash	18.8%
	Limestone	17.6%
		100.0%
Empirical formula (glazes)	Na_2O 0.5 SiO_2 3.0	
	CaO 0.5	
Ionic formula	$Na_{0.33}$ $Ca_{0.17}$ Si $O_{2.3}$	
	or	
	$Na_{0.14}$ $Ca_{0.07}$ $Si_{0.43}$ O	
	m = 0.21 glass network modifier	n = 0.43 glass network former

Methods of expressing glass composition. The student may well become confused by the many methods employed in the literature on the subject to express glass composition. Pincus brings this out clearly and his examples are assembled in Table 17–2 to give a direct comparison.

Single component glasses. The most important of the single oxide glasses is silica. This consists of a three-dimensional network of silicon-oxygen tetrahedrons bonded at every corner, so that each silicon is bonded to four oxygens and every oxygen is bonded to two silicons, giving an ionic expression of $Si_{0.50}O$. Because of the complete and strong bonding, silica glass has a high softening point, high viscosity, low coefficient of expansion, and chemical inertness.

Boric oxide glass, B_2O_3 or $B_{0.67}O$, has a weaker structure than SiO_2, since B^{+++} can surround itself with only three oxygens in triangular coordination. Thus the network approaches a plane, a type with weak forces in one direction. Hence boric oxide glass has a low softening point, is soluble in water, and has a high expansion coefficient. On the other hand, it is a very stable glass with regard to devitrification.

Phosphorous pentoxide glass, P_2O_5 or $P_{0.40}O$, forms tetrahedra with oxygen in the same way as silica, but since there are four O^{--} to one P^{5+}, only three unsaturated valence bonds are available. Thus one oxygen out of the four is linked to only one cation. This means that only three corners of the tetrahedrons are joined, which accounts for the low softening point and hydroscopic properties of this glass. Like B_2O_3 glass, it does not readily devitrify.

Binary glasses. Sodium silicate glasses are well known in the range $Na_2O \cdot 3SiO_2$ ($Na_{0.29}Si_{0.43}O$) to $Na_2O \cdot SiO_2$ ($Na_{0.67}Si_{0.33}O$). As the silica decreases, the tendency to devitrify increases. The Na^+ satisfies the extra O^{--} bonds in the network, but the forces are weak and consequently these glasses are soluble and have low softening points.

The calcium or other alkaline earth silicate glasses are very interesting as they separate into two immiscible liquids. One of these is nearly pure silica while the other, lime-silica, is composed of $71SiO_2$, $29CaO$ ($Ca_{0.17}Si_{0.41}O$). Binary glasses of silica with either lead or zinc oxide occur over much wider ranges of composition than would be expected, and this is explained by the polarizing ability of Pb^{++} and Zn^{++}. P_2O_5 forms glasses readily with the alkaline earths or Al_2O_3.

Ternary glasses. The most important glass is, of course, soda-lime-silica, which forms glass over a considerable range. However, the useful glasses are confined to a small field about Na_2O-CaO-$6SiO_2$.

Elastic and viscous forces in the glass network

We have given in the preceding sections of this chapter something of the atomic arrangement in the glass structure. Now it will be instructive to examine glass in another way, that is, by a study of the integrated forces acting in the glass network.

Thermal expansion. Figure 17–1 shows a linear thermal expansion curve of a normal soda-lime glass. The curve (1) represents an annealed specimen, which expands at a uniform rate until region (a) is reached, where the coefficient increases by a factor of 2 or 3. This zone of inflection is known as the transformation point. As the temperature goes higher, point (b) is reached, which is known as the softening point. This expansion is completely reversible below point (a).

There have been many theories propounded to account for this change in slope at the transformation point (it should really be called a region, since the inflection is never sharp and varies with the rate of heating). It has been shown in the previous sections that glasses with weak bonds have higher expansion coefficients than stronger bonded networks. For example, B_2O_3 glass has thirty times the expansion coefficient of SiO_2 glass. It would seem reasonable to believe that as the temperature is increased,

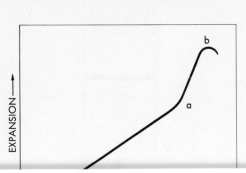

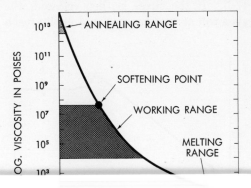

thermal agitation weakens some of the bonds when a certain energy level is reached and thus causes a higher expansion coefficient.

Viscosity. The viscosity of glass is an important characteristic in working and annealing, and therefore its measurement is of practical as well as theoretical importance. The curve of Fig. 17–2 shows the viscosity of a normal soda-lime glass over a wide range of temperature. On the viscosity scale are noted several regions of importance to the glass maker:

Annealing range	$10^{12.5}$–$10^{13.4}$ poises
Softening point	$10^{7.6}$
Working range	10^{4}–$10^{7.6}$
Melting range	$10^{1.5}$–$10^{2.5}$

Extrapolating the viscosity curve to room temperature gives a value of 10^{27} poises. Undoubtedly stressed glass will flow at room temperature, but at a rate so slow that a lifetime would be needed to measure it.

Flow properties. When stress is applied to a glass heated to the annealing range, it will flow somewhat rapidly at first and then at a slower but steady rate as shown in Fig. 17–3. If the stress is then released, the glass will at once contract and finally reach a constant length, but greater than originally. If the straight line is extended to zero time, the intercept *a* will be the elastic flow, and if the same is done when the stress is released, the elastic return *b* will be equal to *a*. It would then appear that the forces in the glass network are of two kinds; first, an elastic force caused by stretching of the stronger bonds, and second, a viscous force caused by a continuous breaking and reforming of the weaker bonds. In other words, there must be discrete groups of atoms that are only distorted elastically, and flow must take place between these groups. This flow may be duplicated by the simple model in Fig. 17–4, where the spring represents the elastic forces and the dash pot the viscous forces.

A glass at constant temperature may be stressed at several values and the uniform flow rate determined from the slope of the

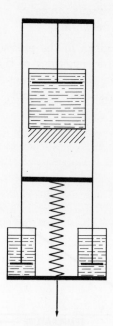

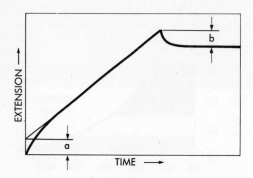

FIG. 17–3. Flow of a stressed glass rod in the annealing range.

FIG. 17–4. Mechanical model to show the flow in glass.

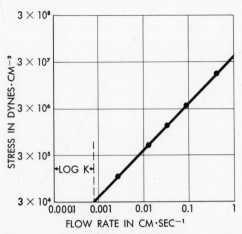

FIG. 17–5. Dependence of flow rate on stress for a heated glass.

curves. If these rates are then plotted against stress on logarithmic paper, a straight line will result as in Fig. 17–5. The slope of this line will be n, for

$$\log \frac{dl}{dt} = \log k + n \log \sigma$$

$$\frac{dl}{dt} = k\sigma^n.$$

When $n = 1$ the flow is viscous, which is the case for glasses.

As shown in Fig. 17–3 the flow rate of glass at constant temperature and stress changes with time, and therefore the measured viscosity changes with time, eventually approaching an equilibrium value from either above or below the final stress value, as shown by Lillie. Looking at it in another way, perhaps the true viscosity does

must be the cooling rate. On the other hand, annealing does more than relieve the cooling stresses; it also permits the atomic structure to settle down into a more stable state as evidenced by higher density and higher index of refraction. In other words, in a quickly cooled glass considerable disorder is frozen in. On annealing, the more compact arrangement is arrived at.

Much thought has been given to the mechanism of annealing. The well-known Adams and Williamson law states that

$$\frac{1}{\sigma} - \frac{1}{\sigma_0} = At,$$

where σ is the stress at time t, σ_0 is the initial stress at zero time, and A is a constant varying with temperature and composition. This equation fits the facts well at the upper end of the annealing range but is not satisfactory at higher viscosities.

Lillie proposes a relation which in many ways gives better correlation with experiment. This is

$$-\frac{d(\log \sigma)}{dt} = \frac{M}{\eta},$$

σ = stress,
t = time,
M = elastic modulus at temperature of annealing,
η = viscosity at temperature of annealing.

Annealing will permit the reduction of stresses to any desired degree if sufficient

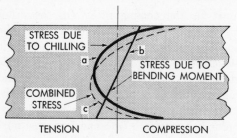

FIG. 17-6. Stresses in a chilled glass sheet.

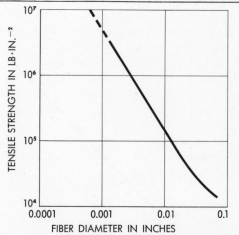

FIG. 17-7. Breaking strength of glass fibers.

brings them at once below the flow temperature, while the center of the sheet is still hot and deformable. The subsequent cooling of the center will then put the surfaces in compression, as shown by curve *a* in Fig. 17-6. Should a sheet annealed in the usual manner be stressed by bending, the forces across the thickness would be as shown in curve *b* and the breaking would occur on the tension side, since glass is much stronger in compression than in tension. If the same load is now applied to the quenched glass, the internal stress distribution is obtained by the sum of the residual stress *a* and the applied stress *b* which gives curve *c*. It will be seen that in this case the tension stress at the surface, which controls the breaking, is changed to compression and the total strength is greatly increased. Another factor in the use of chilled glass is the type of break; instead of the usual jagged fragments it breaks into small cubes.

Properties of glass

There is not space here to do more than touch briefly on this subject. For further information, the reader is referred to that excellent treatise, *The Properties of Glass*, by G. W. Morey.

Mechanical properties. Glass is a material of considerable strength, but compared to metals, it is a brittle material. The theoretical strength of glass may be computed by the bond strengths, and for soda-lime glass amounts to over a million lbs/sq in., a value many times the actual figure. From this it has been deduced that glass contains numerous submicroscopic flaws in the atomic network, and that failure occurs at the weak points caused by these flaws. This is confirmed by the much quoted work of Griffith, wherein he measured the tensile strength of silica glass fibers over a wide range of sizes, the results of which are plotted in Fig. 17-7. It will be seen that as the fiber becomes smaller, there is less probability of its containing a serious flaw and its strength climbs towards the theoretical value. It has also been noted that aged fibers are materially weaker than freshly drawn fibers, due, it is believed, to the hydration of the surface which accentuates the flaws. This subject of flaws is receiving considerable study, since the possibility is at hand for producing glass of greatly increased strength.

The composition of glass influences its mechanical strength to some extent. The high softening, strongly bonded glasses like SiO_2 have the greatest strength, and this decreases as the bonding is weakened.

Optical properties. If glass were not a transparent material, its use would be restricted to a very small field. However, the optical properties comprise much more than transparency, for refraction and dispersion are of particular interest to the lens designer. In the next chapter the influence of composition on these properties will be touched upon.

Electrical properties. The electrical properties of glass must be considered both as to the interior of the structure and as to the surface. In certain cases, such as using glass fibers for electrical insulation, alkali-free compositions are necessary to prevent hydration of the enormous surface area exposed and consequent increased surface leakage.

Thermal properties. The thermal properties of many glasses are well known. There is great need for a glass with a higher use limit than silica glass (1000°C), but as yet nothing superior has been found. Low coefficients of expansion are of value in glass that is subjected to sudden temperature changes. Research here has led to the development of the pyrex types with high silica and low alkali content.

Chemical properties. In general, glass is an excellent container material for chemicals. In fact, the only substance that attacks it rapidly is hydrofluoric acid. Nevertheless, a minute amount of the glass goes into solution even in distilled water, because of a surface leaching of the alkalis. Therefore, the low alkali glasses, such as pyrex, are more suitable for chemical use than is ordinary glass.

REFERENCES

BAIR, G. J., "The Constitution of Lead Oxide-Silica Glasses: I, Atomic Arrangement." *J. Am. Ceram. Soc.* **19**, 339, 1936.

KREIDL, N., and WEYL, W. A., "The Development of Low Melting Glasses on the Basis of Structural Considerations." *Glass Industry*, I, 335; II, 384; III, 426; IV, 465, 1941.

MOREY, G. W., *The Properties of Glass.* Reinhold Publishing Corp., New York, 1938.

STANWORTH, J. E., *Physical Properties of Glass.* Oxford University Press, Toronto, 1950.

SUN, K., "Fundamental Condition of Glass Formation." *J. Am. Ceram. Soc.* **30**, 277, 1947.

WARREN, B. E., "X-ray Diffraction of Vitreous Silica." *Z. Krist.* **86**, 349, 1933.

ZACHARIASEN, W. H., "The Atomic Arrangement in Glass." *J. Am. Chem. Soc.*, 3841, 1932.

CHAPTER 18

GLASSES

Introduction

The glass industry amounts to approximately one-third of the whole ceramic industry in value of annual production. The glass producers are progressive in research and development as well as in engineering, a fact that has done much to raise this industry to its present level.

Glass compositions

Commercial glasses. The great bulk of the glass produced today is of the soda-lime-silica type with a small addition of alumina. This glass is used for containers and sheet glass. Figure 18–1 shows the trends in composition of this type of glass for the last few years. Increasing the alumina content adds to the chemical resistance and in some ways aids the working properties. In Table 18–1 is given a typical analysis of a container glass. It is obvious that this composition may be produced with a large variety of raw material. Cost and availability would suggest silica sand, limestone, and soda ash, which are indeed the main ingredients, but it has been found desirable to add some of the soda in the form of the sulphate or nitrate, a procedure which is explained later in this chapter, and to add feldspar or nepheline syenite as a source of alumina. While sheet glass is not decolorized, much of the container glass uses a decolorizer as discussed in Chapter 21.

Glass for tableware is quite similar to the container glass, as shown in Table 18–1, except that it is lower in iron content and often contains barium oxide for greater brilliance.

The so-called flint glass or crystal is used for the highest grade of artware, especially when it is to be cut or engraved. This glass contains a considerable proportion of lead oxide, as shown in Table 18–1, which increases the index of refraction and thereby gives greater brilliance. The lead also makes the glass softer for cutting.

High silica glasses of the pyrex type have the advantages of low thermal expansion, high softening point, and good chemical resistance. Therefore they are much used for

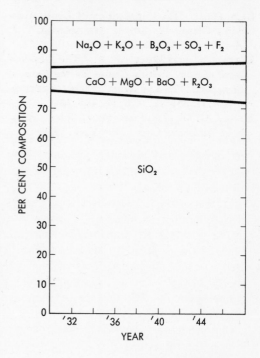

FIG. 18–1. Change in glass composition.

156

Table 18-1

Glass Compositions

Constituent	Type of glass									
	Window	Container	Pyrex	Textile fiber	Opal	Light flint	Heavy flint	Barium crown	Crown	Thermometer glass
SiO_2	72.1	72.1	80.5	54.0	65.8	67.4	46.1	59.1	72.2	67.5
B_2O_3			11.8	10.0				3.0	5.9	2.0
Al_2O_3	1.1	1.8	2.0	14.0	6.6	1.7	0.1	0.1		2.5
Fe_2O_3	0.2	0.1								
As_2O_3			0.7				0.4	0.3	0.2	
ZnO						3.9		5.0		7.0
CaO	10.2	5.6	0.3	17.5	10.1	0.4	0.1	0.1	2.1	7.0
MgO	2.6	4.2	0.1	4.5					0.1	
BaO		0.3						19.3		
PbO						10.7	45.1			
K_2O			0.2		9.6	0.1	6.8	9.7	13.9	
Na_2O	13.6	15.6	4.4		3.8	15.1	1.7	3.2	5.2	14.0
Sb_2O_3									0.1	
SO_3							0.1		0.1	
F_2					5.3					

laboratory and cooking ware. The general composition as shown in Table 18-1 indicates a low alkali content, the virtual elimination of alkaline earths, and the use of boric oxide to achieve a reasonable temperature for melting and working, although these temperatures are considerably higher than for container glass.

Fiber glass of the textile variety, because of its enormous surface area, must be particularly stable to atmospheric moisture. Therefore a completely alkali-free glass is used for this purpose, as shown in Table 18-1. This glass has a rather high softening temperature, but also has viscosity properties that permit drawing fibers.

Optical glasses. There is not space here to discuss the thousands of glass compositions designed for optical use; however, a few typical compositions are shown in Table 18-1.

The lens designer desires glasses not only with a considerable range of index of refraction and dispersion, but the greatest possible range in the ratio of the two. To show this picture, we cannot do better than to examine the excellent chart from Pincus (Fig. 18-2) showing areas of the various glass types. Tremendous strides have been made in optical glasses during and since World War II, as shown by the added fields in the diagrams. The use of fluorides and rare elements has produced many excellent glasses. This work has been accomplished largely by the understanding of the principles of crystal chemistry, and great credit must go to the Eastman Kodak Company for much of this development. With these new glasses, it is possible to design simpler and more highly corrected photographic objectives than were previously possible.

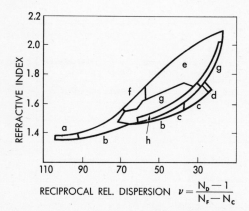

Table 18-2

Typical Container Glass Batch

Sand	473
Limestone (burned)	186
Soda ash	165
Salt cake	18
Arsenic trioxide	3
Cullet	151
Manganese dioxide	5

FIG. 18–2. Optical glasses: (a) Eastman Kodak Fluoride; (b) Eastman Kodak Fluophosphate; (c) Eastman Kodak Fluorosilicate (d) Eastman Kodak Fluogermanate; (e) Eastman Kodak Rare element borate; (f) Eastman Kodak Fluoborate; (g) glasses prior to 1934; (h) glasses prior to 1880.

Special glasses. There are many special glasses used, but here only a few can be mentioned. One of these is opal glass, used for containers and light diffusers. The production of opaque and translucent glasses will be discussed in Chapter 19. A commercial opal as shown in Table 18–1 contains fluorine, which produces crystals on cooling.

The various colored glasses will be taken up in Chapter 21; in this field large production is confined principally to green and amber for beverage bottles.

Mechanism of melting

The melting process is an interesting one, since it consists of many types of thermochemical processes. In this discussion we will confine ourselves to a typical container glass. It should be remembered that the object of melting is to convert the batch materials into a homogeneous glass of constant properties for the forming operation, in the most economical manner.

Batch. A typical batch for container glass is shown in Table 18–2. It will be seen that besides the main ingredients of sand, limestone, and soda ash, there are several other materials. The feldspar is a source of alumina, while at the same time it adds alkalis and silica. The sodium nitrate, of course, adds soda to the glass, but it also acts as an oxidizing agent when it decomposes. The arsenic is a fining agent, giving off oxygen at the proper temperature to form large bubbles which rise through the glass and sweep out the small bubbles or seeds, as they are called. The manganese is a physical decolorizer that produces a pink color complementary to the green of ferrous iron. Iron is largely in the reduced state, even with an oxidizing agent, since the equilibrium goes in the direction of FeO at high temperatures.

Melting process. The melting process is an involved series of reactions that take place as the temperature rises and continue with time after the temperature has reached its maximum. It is believed that the best way to portray them is by a chart, shown in Fig. 18–3, where the reactions are related to the temperature. It should be realized that most of these reactions occur simultaneously, but as yet we have no reliable data on the exact life history from the raw batch to the refined glass. Figure 18–4 shows a more or less imaginary chart depicting the melting process. It should not be too difficult to obtain the data to make this chart real.

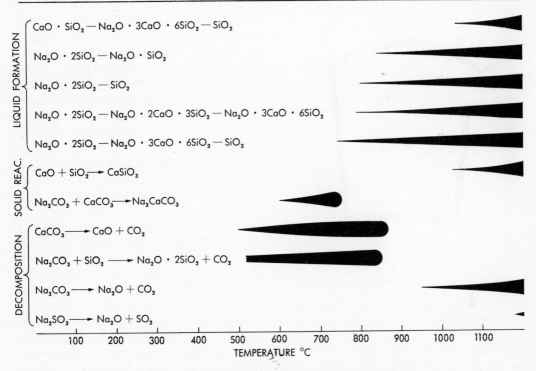

FIG. 18–3. Some of the reactions occurring in glass melting.

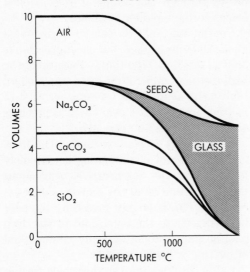

FIG. 18–4. Progress of melting of the glass batch.

Melting equipment. Glass used to be melted in pots made from special clays. In this process, since heat is applied largely from the outside of the pot, the pot wall must be as hot as or hotter than the glass, a fact that makes necessary a pot refractory with good hot strength to withstand the hydrostatic pressure of the glass. Also the high refractory temperature offers an opportunity for considerable reaction between the glass and refractory.

At the present time pots are still used for melting special glasses where the amount does not justify the use of a tank, such as in the case of optical glasses, colored glasses, or flint glass for cutting, and for some plate glass. Figure 18–5 shows a cross section of a typical covered pot which is used in a multiple pot furnace, a type that has changed little in the last 125 years. Pots range in

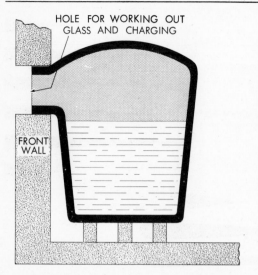

HOLE FOR WORKING OUT GLASS AND CHARGING

FRONT WALL

FIG. 18-5. Cross section of a glass pot.

size from those holding 500 lb of glass up to those holding 4000 lb, and the time from filling to working may be as little as 24 hours, at temperatures of around 1400°C. The life of the pot is several months with periodic charging and working out. As the whole furnace cannot be cooled down to change one pot, it is necessary to preheat the new pot, break down a portion of the furnace wall, remove the white-hot old pot, and replace it with the new one — a hot and exacting operation.

A considerable amount of optical glass is now melted in platinum-lined pots. While the platinum is very costly, it may be reworked with a loss of less than 10 per cent of its value. The glass produced in platinum pots is especially homogeneous, since there is no solution of refractory as in the case of clay pots.

Obviously, the cost of melting in pots is high, so a less expensive method for larger quantities was developed in the form of a day tank. This melts glass in batches like the pot, but there is a fundamental difference. Here the glass is heated entirely from

the free surface, and the refractory temperature is always lower than the mean glass temperature. This allows less reaction between the glass and refractory but causes large temperature gradients in the glass and tends to produce inhomogeneity. Glasses more transparent to infrared radiations shows more even temperatures, as might be expected.

When still larger quantities of glass are needed to supply automatic forming machinery, the continuous tank is used. This is similar to the day tank, except that the batch is continuously fed into one end and glass is continuously drawn out of the other end. In Fig. 18-6 is shown a section of a glass tank for container glass. Heat is supplied to the melting section where the batch is fused. The resulting glass then flows through a submerged opening, the *throat*, to remove the scum, and moves slowly at a lower temperature through the refining section, where the fining is completed. It then passes through the temperature-controlled feeder to the molding machines.

Continuous tanks for container glass hold about 200 tons of glass, but sheet glass tanks hold as much as 1400 tons. Because of the high melting temperatures used, the modern tank melts as much as 0.2 tons per sq ft of melting area per day, over twice as much as was the practice 30 years ago.

Heat balance in glass melting. The thermal efficiency of glass melting is extremely low even in the regenerative, continuous tank. A value of 20 per cent is seldom exceeded. This is largely caused by the poor heat transfer to the batch and to the deep pool of glass. Some radically new method of melting is badly needed.

In Fig. 18-7 is shown the heat balance diagram of a typical continuous tank. The use of fusion cast refractories has greatly lengthened the tank life, but at the expense of the

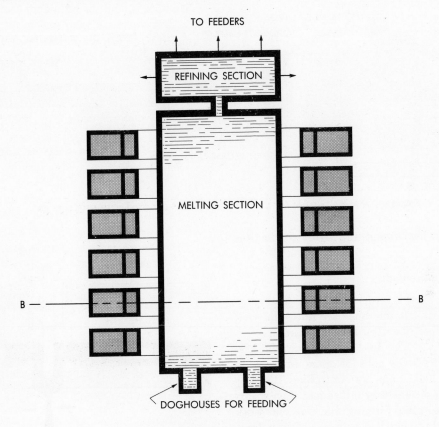

HORIZONTAL SECTION ON AA

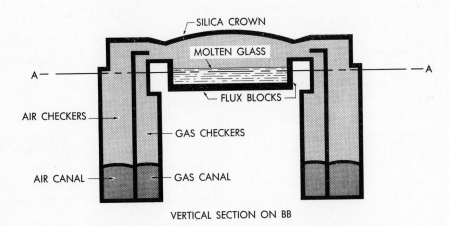

VERTICAL SECTION ON BB

Fig. 18–6. Tank for melting container glass.

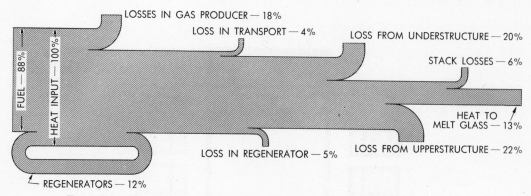

FIG. 18–7. Heat balance for a continuous glass tank fired with producer gas.

heat flowing through the highly conductible blocks.

Forming methods

Since glass is formed while it is in the viscous state, the methods are quite different from those described for clay ware.

Feeders. The feeder, as shown in Fig. 18–8, produces gobs of glass uniform in weight and shape by controlling the flow through a refractory orifice and then shearing. No glass forming machine can work efficiently unless it is supplied with uniform glass.

Pressing. This method is used for plates, tapered tumblers, and other thick-walled objects. It is believed that the first glass press was developed by a carpenter at the Sandwich Glass Company in the year 1816. This was the forerunner of the automatic glass forming machine which later revolutionized the industry.

Figure 18–9 shows the steps in pressing a dish. The molds are made of a special cast iron surfaced on the inside with a firm, smooth layer of carbon developed from oily coatings. A typical coating medium is made up as follows: linseed oil, 1000 gm; rosin, 200 gm; and fine cork dust, 100 gm. The mold must be kept at the correct tem-

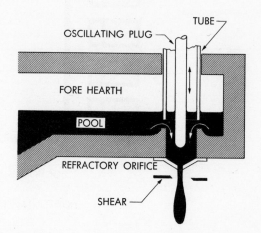

FIG. 18–8. Cross section of a gob feeder.

perature, for if it is too hot, the glass will stick and if too cold, it will not form an even surface.

Blowing. Blowing is used for most containers, such as bottles and jars. There are several processes used, but the ones illustrated in Figs. 18–10, 18–11, and 18–12 show the principles of forming clearly. It will be seen that there are two steps in each method, the first a forming of the parison or temporary shape from the gob, and the second the blowing of the parison to fit the inside of the blow mold. While the first

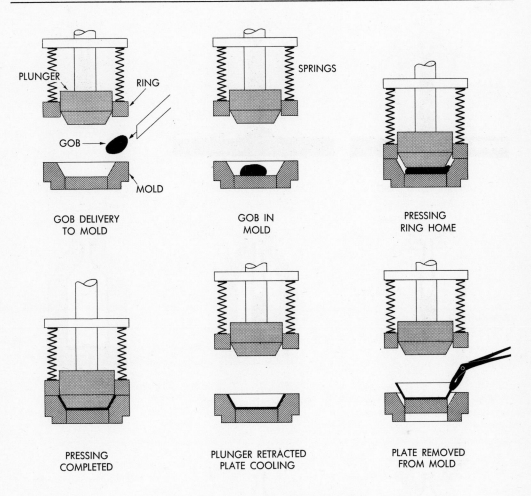

PLUNGER
RING
SPRINGS
GOB
MOLD

GOB DELIVERY
TO MOLD

GOB IN
MOLD

PRESSING
RING HOME

PRESSING
COMPLETED

PLUNGER RETRACTED
PLATE COOLING

PLATE REMOVED
FROM MOLD

FIG. 18–9. Steps in pressing a glass dish.

step varies with the different methods, the second step is the same in all of them. During the blowing, the thicker portions of glass retain their heat longer than the thin portions and therefore flow more, thus producing a more even wall thickness.

The molds here are special cast iron maintained at a temperature of about 400°F and coated with a mold paste. The machines that operate the molds are fully automatic, but will not be described here.

Drawing. Sheet glass is now formed in this country by drawing a continuous sheet from a pool in the feeder of the glass tank, as shown in Fig. 18–13. The maintenance of exact temperatures in the sheet at each point in the cycle is very important. Some sheet is rolled as shown in Fig. 18–14, the glass running over a refractory weir. Tubing or rods may also be drawn, either intermittently or continuously.

Spinning. The use of glass in the form

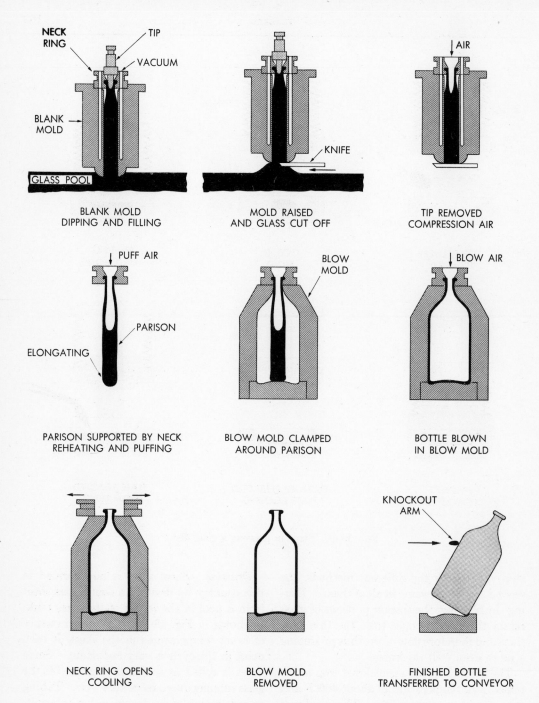

FIG. 18–10. Vacuum and blowing process (Owens Machine).

164

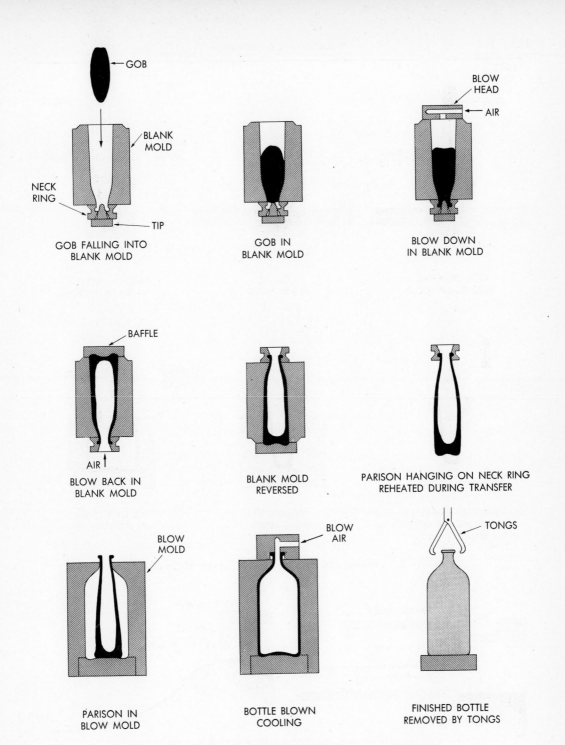

Fig. 18–11. Blowing and blowing process (Lynch Machine).

165

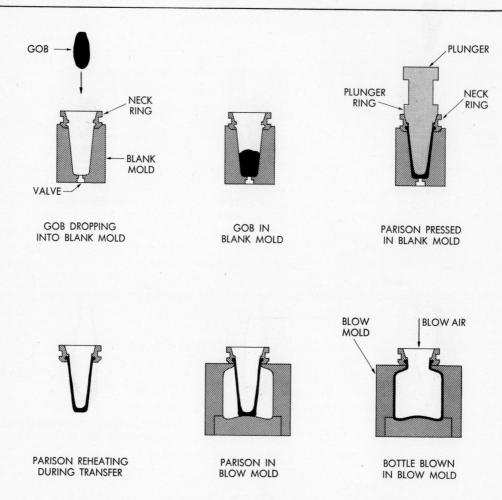

FIG. 18–12. Pressing and blowing process (Lynch Miller Machines).

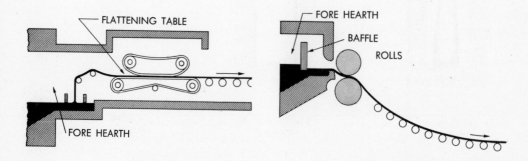

FIG. 18–13. Drawing sheet glass.

FIG. 18–14. Rolling sheet glass.

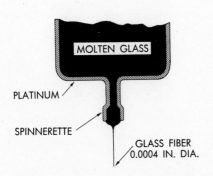

FIG. 18–15. Cross section of a spinnerette used in forming glass textile fibers.

of fibers has been very extensive. This material may be divided into two types; one is the continuous filament for textile use and the other the discontinuous fiber employed in heat insulation, filters and plastic reenforcement.

The continuous filaments are drawn from a series, often 204, of platinum spinnerettes set in the bottom of a platinum heating chamber. The glass is fed into this chamber in the form of glass marbles at a rate that keeps the glass level constant. A cross section of a spinnerette and fiber are shown in Fig. 18–15. The fiber is pulled from the bottom of a drop of glass held in the mouth of the spinnerette and the flow of glass through the passage of about 0.07 in. diameter simply replenishes the drop. As the viscosity of the glass must be close to the correct value, the temperature control must be exact. The temperature is maintained by electric resistance heating of the platinum chamber, although one manufacturer uses high frequency induction heating that has the advantage of greater uniformity.

The bundle of fibers is reeled up at the rate of about 10,000 ft per minute and the diameter ranges between 0.0002 and 0.0005 in. If fibers break or are broken when changing reels, they are readily started again, since the drop on the spinnerette falls by gravity and draws a new fiber behind it.

Discontinuous fibers are blown by striking a stream of glass or slag with a high velocity jet of steam. The rock wool is made as shown in Fig. 18–16, where the high vorticity of the moving steam elongates the fiber. All blown fiber contains some shot, 15 to 25 per cent by weight, and many believe that these shot are formed first and are then drawn out into fibers as they move through the air. This is not always so; the fibers are often formed first, as is clearly shown in the photographs, and if not chilled rapidly enough, surface tension draws them up into spheres. Glass fibers are blown from a multiple stream of glass running from a series of nozzles in the bottom of an electrically heated platinum melting chamber. In this case the air or steam blast is parallel to the glass stream and accelerates it to a very high velocity.

Some fibers are produced by having the glass or melted rock stream fall on a rapidly revolving disk; this produces a low short-fiber content.

Finishing and annealing

Fire polishing. Laboratory ware or tableware, after forming, is usually smoothed on the seams and edges by applying localized heat. This lowers the viscosity to the point where surface tension forces can level the surface.

Grinding. Light tableware and artware is often finished by grinding and polishing the edges on a wheel.

Annealing. After finishing, the glassware is cooled at a given rate in a *lehr* (annealing oven). Usually this is a continuous process with the ware passing through the heated chamber on a moving mesh belt. Sheet

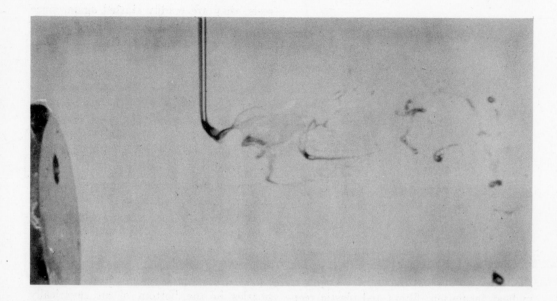

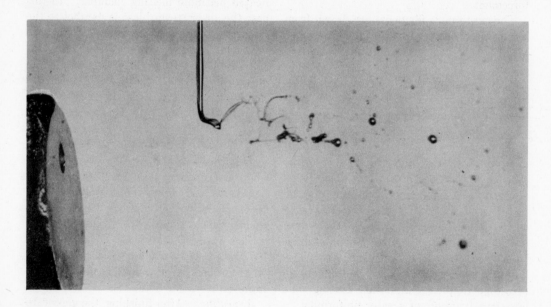

Fig. 18–16.   Molten rosin blown into fibers to simulate slag wool production.   The photographic exposure was only three microseconds.   In the upper picture, normal rosin temperature forms good fibers, while in the lower one a higher temperature produces much shot.

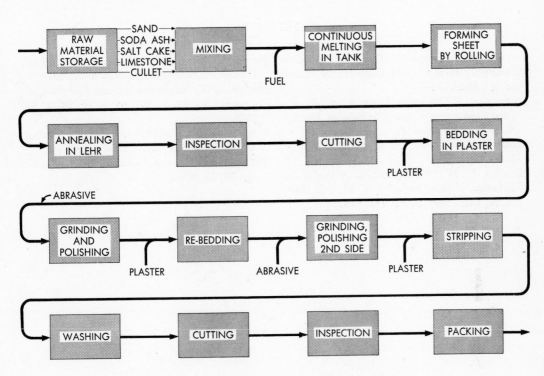

FIG. 18–17.   Flow sheet for the production of plate glass.

glass is handled in the same way with the sheet moving at the drawing speed through a long lehr.

Grinding and polishing plate glass

Grinding.   This operation takes place on the face of large plates that are bedded down in plaster.   Wet sand is used as an abrasive under revolving cast-iron laps.   At first the sand is quite coarse, but as the grinding proceeds it becomes finer and finer until a very smooth, mat surface is produced on the glass. In a modern plant, as shown in the flow sheet of Fig. 18–17, the operation is continuous; the sheets of glass mounted on cars move under a series of laps, each of which is fed with the proper size of sand.

Polishing.   The finely ground sheet is next polished by passing it under felt laps fed with a suspension of fine rouge (Fe_2O_3) in water.   The mechanism of polishing glass is not completely understood, but it is generally believed that a thin layer of the surface actually flows due to the high temperatures and pressures.   Thus the humps are slid into the hollows to produce a level surface.   Some think the glass is actually melted and that only refractory particles are capable of polishing.   However, there is no known explanation of why rouge and cerium oxide are among the very few efficient polishing agents.

Optical glass is not only polished with felt laps but also with hard ones made from a special pitch.   Care must be taken to eliminate all oversized particles in the suspension to prevent scratching.

REFERENCES

DAVIS, P., *The Development of the American Glass Industry*. Harvard University Press, Cambridge, Mass., 1949.

DEVILLERS, R. W., and VAEREWYCK, F. E., *Glass Tank Furnaces*, trans. by Scholes. Ogden-Whitney Press, New York, 1937.

"Glass Glossary." *Bull. Am. Ceram. Soc.* **27**, 355, 1948.

HOWARD, G. E., "Mechanization of the Glass Industry." *Glass Industry*, I, **31**, 75, 1950; II, **31**, 131, 1950; III, **31**, 183, 1950.

PHILLIPS, C. J., *Glass the Miracle Maker*, 2nd ed. Pitman Publishing Corp., New York, 1948.

PINCUS, A. G., "Invitation to Glass Technology." *Ceramic Age* **53**, 5, 260; **54**, 1, 19; **54**, 2, 81, 1949.

SCHOLES, S. R., *Modern Glass Practice*. Industrial Publications, Inc., Chicago, 1941.

WEYL, W. A., "Glass Composition." *Ceramic Age* **50**, 54, 1947.

CHAPTER 19

GLAZES

Introduction

A glaze may be defined as a continuous adherent layer of glass, or glass and crystals, on the surface of a ceramic body. The glaze is usually applied as a suspension of the glaze-forming ingredients in water, which dries on the surface of the piece in a layer. On firing, the ingredients react and melt to form a thin layer of glass. The glaze may be fired at the same time as the body or in a second firing.

The main purpose of the glaze is to provide a surface that is hard, nonabsorbent, and easily cleaned. At the same time the glaze permits the attainment of a greater variety of surface colors and textures than would be possible with the body alone.

Methods of expressing glaze compositions

Batch formula. The glaze may be specified by giving the weights of the ingredients. For example, a typical raw lead glaze is:

White lead	154.8 gm
Whiting	30.0
Feldspar	55.7
Kaolin	25.8
Flint	48.0
	———
Total	314.3

This list of ingredient weights is excellent for the person mixing the glaze, but it does not permit the technologist to visualize the nature of the glass structure formed, or to compare the glaze with another.

Equivalent formula. The glaze given in the previous section may be expressed in

molecular equivalents of the oxides, as first suggested to the ceramists by Segar, which gives:

$$0.6 \text{ PbO}$$
$$0.3 \text{ CaO} \qquad 0.2 \text{ Al}_2\text{O}_3 \qquad 1.6 \text{ SiO}_2$$
$$0.1 \text{ K}_2\text{O}$$

The RO or basic oxides (where R is any cation) are placed in the first column with a total of unity, the R_2O_3 or amphoteric oxides in the second column, and the RO_2 or acid oxides in the last column. While this method of expressing glazes was developed long ago, it can be seen in the light of our modern knowledge of crystal chemistry that in general the constituents have been grouped respectively into glass network modifiers, intermediates, and formers.

Ionic formula. As shown in Chapter 17, a glass or glaze may be expressed in the ionic form with the network formers as unity or the total anions as unity. Thus our raw lead glaze becomes:

$$\text{Pb}_{0.6}^{++} \quad \text{Ca}_{0.3}^{++} \quad \text{K}_{0.2}^{+} \quad \text{Al}_{0.4}^{+++} \quad \text{Si}_{1.6}^{++++} \quad \text{O}_{4.8}^{--},$$

or

$$\text{Pb}_{0.38}^{++} \quad \text{Ca}_{0.19}^{++} \quad \text{K}_{0.13}^{+} \quad \text{Al}_{0.25}^{+++} \quad \text{Si}^{++++} \quad \text{O}_{3.0}^{--}.$$

The silicon-oxygen ratio then is 0.33. This is of the proper magnitude for a stable glass, for if the anions are unity, we get

$$\text{Pb}_{0.13}^{++} \quad \text{Ca}_{0.06}^{++} \quad \text{K}_{0.04}^{+} \quad \text{Al}_{0.08}^{+++} \quad \text{Si}_{0.33}^{+++} \quad \text{O}^{--}$$

Then $m = 0.31$ and $n = 0.33$, as shown in Table 17–2.

Classification of glazes as to composition. The following list shows the common types of glazes.

1. Raw glazes (containing insoluble raw materials)
    a. lead-containing glazes
    b. zinc-containing glazes (Bristol)
    c. porcelain glazes
2. Fritted glazes (containing some glass before firing)
    a. lead-containing glazes
    b. leadless glazes
3. Vapor glazes (deposited from the vapor phase)
    a. salt glaze
    b. smear glaze

The composition ranges of these glazes conform in a general way to the chart of Fig. 19–1, redrawn from one by Holscher and Watts.

Classification of glazes as to surface. It is also possible to classify glazes according to surface characteristics, thus:

> Glossy
> Semi-mat
> Mat
> Surface crystalline
> Vellum

Classification of glazes as to optical properties. Also they may be classified according to the nature of the interior of the glaze layer, as:

> Transparent (clear)
> Opaque (Majolica, enamel)
> Fine interior crystals (Aventurine)
> Large interior crystals (crystalline)

Of course, glazes may be readily classified as to color.

Methods of compounding glazes

Raw glazes. In ceramic literature, glazes are often given in the equivalent form, so that it is necessary to change this to the

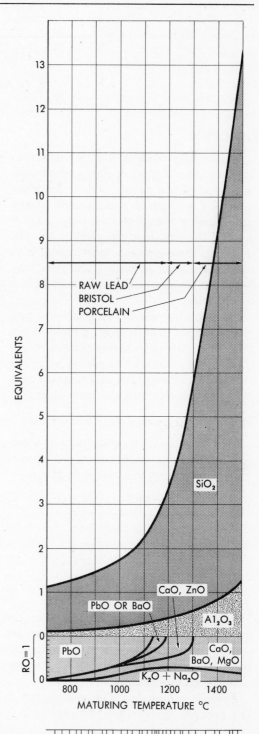

Fig. 19–1. Graphic representation of raw glazes. For example, a raw lead glaze maturing at 1100°C is: 0.3 PbO, 0.25 K_2O, 0.45 CaO, 0.3 Al_2O_3, 2 SiO_2.

batch formula before the glaze can be made up. Anyone even without a knowledge of chemistry will find this quite simple, since it is really nothing but an arithmetical operation.

First, the term *equivalent weight* should be made clear. For example, potash feldspar, $K_2O \cdot Al_2O_3 \cdot 6SiO_2$ has a formula weight of 556.8 gm as shown in Table A-11 in the appendix. Thus 556.8 gm of potash feldspar will yield one molecular equivalent (94.2 gm) of K_2O, one molecular equivalent (101.9 gm) of Al_2O_3, but six molecular equivalents (360.6 gm) of SiO_2. Therefore, if it is desired to add one equivalent of SiO_2 by means of feldspar, not 556.8 gm but rather 556.8/6 or 92.8 gm of it would be added.

As an example take a simple raw lead glaze such as:

$$0.1 \ K_2O$$
$$0.8 \ PbO \qquad 0.2 \ Al_2O_3 \qquad 3.3 \ SiO_2$$
$$0.1 \ CaO$$

In converting to a batch formula some judgment must be exercised in choosing the raw materials. One could derive K_2O from K_2CO_3, but as this is soluble it would not be suitable for a raw glaze. Thus feldspar is the logical material. In the same way CaO could be obtained from several chemicals, but the carbonate is the most convenient. The PbO could come from red lead or white lead, but the latter is generally preferred as it stays in suspension better. The kaolin is a suspending medium, but not more than 0.15 equivalent should be added in the raw state or too much shrinkage will occur in drying. If more is needed to supply Al_2O_3, a portion of the kaolin should be calcined.

The student should get in the habit of carrying out the glaze calculation in a systematic manner, as shown in Table 19-1. The equivalents of each ingredient are now multiplied by their respective equivalent weights to give the batch weights or percentage composition, as shown in Table 19-2.

The conversion of the batch formula back to the empirical formula is carried out by reversing the process, but in this case no judgment is required in selecting ingredients. For any exact calculation, the actual composition of the materials would have to be considered. For example, kaolin often contains some quartz and feldspar, and feldspar contains some quartz.

Fritted glazes. The calculation of fritted glazes is more complicated than that of raw glazes. The main purpose of fritting is to be able to use water soluble materials by melting them together with other materials to form a relatively insoluble glass. For example, boron compounds are nearly all soluble and must be made into an insoluble borate glass before being used in the glaze suspension. Secondary purposes of fritting are to obtain better working properties of the wet glaze, to distribute color more evenly, or to handle the lead in a less poisonous form.

In compounding a frit there are certain rules usually followed in selecting the composition so that the melted frit will form a glass, fluid enough to flow and at the same time sufficiently insoluble to be ground in water.

These rules are:

1. The ratio of the basic oxides to the acid oxides should be between 1–1 and 1–3. This means that the composition must be in the glass-forming range. If the B_2O_3 content is high, the ratio may be much larger.
2. All soluble alkalis and boric oxide should be in the frit.
3. The ratio of the alkaline oxides to the other basic oxides should not be much more than one.
4. The ratio of B_2O_3 to SiO_2 should be less than one-half.
5. The alumina content should be less than 0.4 equivalent.

Table 19-1

Original glaze	0.1 K_2O	0.1 CaO	0.8 PbO	0.2 Al_2O_3	3.0 SiO_2
0.1 Potash feldspar	0.1 K_2O	---	---	0.1 Al_2O_3	0.6 SiO_2
Remainder	0	0.1 CaO	0.8 PbO	0.1 Al_2O_3	2.4 SiO_2
0.1 Whiting		0.1 CaO	---	---	---
Remainder		0	0.8 PbO	0.1 Al_2O_3	2.4 SiO_2
0.8 White lead			0.8 PbO	---	---
Remainder			0	0.1 Al_2O_3	2.4 SiO_2
0.1 Kaolin				0.1 Al_2O_3	0.2 SiO_2
Remainder				0	2.2 SiO_2
2.2 Flint					2.2 SiO_2
Remainder					0

Table 19-2

Equivalent	Ingredient	Approximate formula	Equivalent weight	Batch weight	Per cent
0.1	Potash feldspar	$K_2O \cdot Al_2O_3 \cdot 6SiO_2$	556.8	55.7	12.0
0.1	Whiting	$CaCO_3$	100.1	10.0	2.2
0.8	White lead	$2PbCO_3 \cdot Pb(OH)_2$	298.3	238.6	51.7
0.1	Kaolin	$Al_2O_3 \cdot 2SiO_2 \cdot 2H_2O$	258.1	25.8	5.6
2.2	Flint	SiO_2	60.1	132.2	28.5
Total				462.3	100.0

Table 19-3

Original form	0.3 Na_2O	0.4 CaO	0.3 BaO	0.1 Al_2O_3	2.6 SiO_2
0.60 Frit	0.3 Na_2O	0.2 CaO	0.1 BaO	---	1.0 SiO_2
Remainder	0	0.2 CaO	0.2 BaO	0.1 Al_2O_3	1.6 SiO_2
0.20 Whiting		0.2 CaO	---	---	---
Remainder		0	0.2 BaO	0.1 Al_2O_3	1.6 SiO_2
0.20 Barium carb.			0.2 BaO	---	---
Remainder			0	0.1 Al_2O_3	1.6 SiO_2
0.10 Kaolin				0.1 Al_2O_3	0.2 SiO_2
Remainder				0	1.4 SiO_2
1.40 Silica					1.4 SiO_2
Remainder					0

Table 19-4

Equivalent	Material	Equivalent weight	Weight	Per cent
0.50	Sodium carb. (cryst.)	286.2	143.1	46.2
0.33	Whiting	100.1	33.0	10.6
0.17	Barium carb.	197.4	33.5	10.8
1.67	Flint	60.1	100.5	32.6
Total			310.1	100.0

Table 19-5

Equivalent	Material	Equivalent weight	Weight	Per cent
0.60	Frit	175.9	105.5	38.4
0.20	Whiting	100.1	20.0	7.3
0.20	Barium carb.	197.4	39.5	14.4
0.10	Kaolin	258.1	25.8	9.4
1.40	Flint	60.1	84.0	30.5
Total			274.8	100.0

As an example, take the following glaze:

$$0.3 \ Na_2O$$
$$0.4 \ CaO \quad 0.1 \ Al_2O_3 \quad 2.6 \ SiO_2$$
$$0.3 \ BaO$$

It is evident that the high Na_2O content will make it impossible to use feldspar without increasing the Al_2O_3 content too much. Therefore, most or all of the soda must come from soda ash (Na_2CO_3). Following the rules previously given, the following frit is reasonable:

$$0.30 \ Na_2O \qquad \qquad 0.50 \ Na_2O$$
$$0.20 \ CaO \quad 1.0 \ SiO_2 \ or \ 0.33 \ CaO \quad 1.67 \ SiO$$
$$0.10 \ BaO \qquad \qquad 0.17 \ BaO$$

The complete glaze is worked out as before (Table 19-3). The batch weights of the frit may be found as shown in Table 19-4. It should be kept in mind that the total weight given above as 310.1 is the weight of the raw materials in the batch, and not the result-

ant weight of frit. Therefore, 0.6×310.1 gm of frit are not added to the glaze but rather 0.6×175.9, calculated on the basis of the oxides alone. Now the complete batch formula can be made up as in Table 19-5.

The frit may be melted in crucibles for small batches, but production lots are made in rotary frit furnaces or small glass tanks. It is quenched in water and ground to about 35 mesh. It can then be mixed with the remainder of the glaze batch and wet-milled for the correct length of time. One thing the glaze formula does not indicate is the fineness of grinding. The working properties of the glaze, as well as the maturing temperature, depend to a considerable extent on the fineness, so the milling conditions must be closely controlled.

Application of glazes

Properties of the glaze suspension. Anyone who has worked with laboratory made glazes and compared them with a carefully developed commercial glaze will see at once the superiority of the latter in working properties. This is due to controlled particle size and correct selection of suspending clays.

The glaze slip should have these properties:

1. Low rate of settling.
2. A high mobility (low viscosity) so that it will flow out into a smooth surface.
3. A high yield point, not in the slip form but after it has lost a little water, so that the glaze layer will not slough off.
4. A low drying shrinkage.
5. A high elasticity in the dry condition.
6. Little change of slip properties with aging.

These conditions are met only with very careful control of specific gravity, particle size, *pH* of the suspension, and type of clay, and perhaps with added organic matter in the form of water soluble gums.

Methods of application. Glazes may be applied by brushing, pouring, dipping, or spraying. The latter method will be discussed here, since it is the principal commercial one.

The glaze suspension may readily be atomized in a spray head with compressed air in the same manner as paint. As the tiny glaze droplets strike the ware, they flatten out and soon lose a portion of their water by absorption in the pores of the body or by evaporation on a vitreous body. Successive droplets build up one on another, but the rate of application must be slow enough so that the water content of the glaze layer does not permit the yield point to be exceeded, and at the same time must be fast enough so that each droplet coalesces into the preceding ones to make a coherent layer.

Spraying is done by hand on large pieces, such as terra cotta, but dinnerware is largely sprayed by automatic machines. One of the problems is the economical recovery of the waste glaze.

Firing the glaze

Life history of a glaze. The diagrams in Fig. 19–2 show the stages of reaction taking place in a simple fritted whiteware glaze — a reaction which is quite similar to the melting of glass except for the reaction with the body. In the first stage the glaze particles frit together and reduce the pore volume. The next step is the formation of a continuous glassy phase with the entrapment of bubbles left from the pores and the decomposition of the carbonates and clay. These bubbles work to the surface and break, where they form pits which soon smooth out. The forces bringing the bubbles to the surface are not gravity forces, as is the case in fining glass, but rather surface tension forces attempting to bring the free surface area to a minimum.

The reaction between the glaze and body is an important one, as it forms an intermediate layer between the properties of the body and those of the glaze itself. This layer often contains mullite needles that grow from the body into the glaze, thus serving as anchors. The intermediate zone is better developed in cases where the body and glaze are fired together, since the interaction can be more complete.

Surface. It is desirable to obtain a smooth surface on a bright glaze. Superficially most bright glazes look smooth, but a careful examination under oblique illumination will show a number of shallow pits. In the case of some low-fired whiteware, the pits are so numerous that the gloss is definitely impaired, whereas the fine porcelain of Copenhagen has an almost perfect surface.

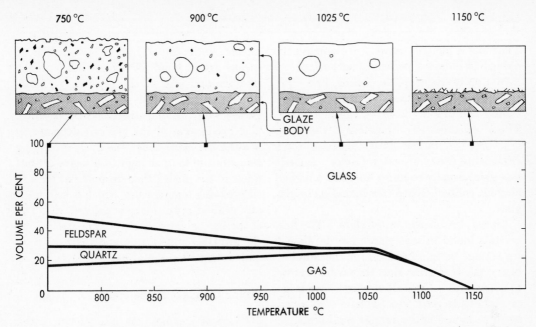

FIG. 19–2. Life history of a fritted glaze. Below are shown the volumes of the various constituents and above, thin sections of the glaze and body at various temperatures.

FIG. 19–3. Willemite crystal grown in a glaze (full size).

177

Time and a nonporous body are apparently needed to obtain a perfect glaze surface.

Mat glazes develop, on cooling, fine crystals in the surface that break up the continuity and produce a sort of eggshell finish. These crystals are commonly anorthite ($CaAl_2Si_4O_{12}$), but mullite ($Si_2Al_6O_{13}$) and cristobalite (SiO_2) are often found. A good mat glaze usually contains lime and a higher alumina content than a corresponding bright glaze.

Crystal development in the glaze. The fine crystals found in mat glazes are developed by normal firing schedules, for the number of crystals is so great that they cannot grow to a large size. Crystalline glazes have large crystals which grow to a diameter of as much as three inches under proper conditions, as shown in Fig. 19–3.

It has been shown that in growing crystals in glass it is first necessary to have a nucleus of a few atom groups as a starting point and then to cause this nucleus to grow. Certain glasses have separate temperature ranges for

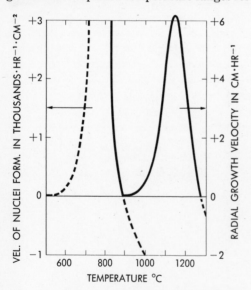

FIG. 19–4. Illustrating the formation of crystals from a glaze.

nuclear formation and for crystal growth, so means are at hand for controlling both the number and size of the crystals. This is illustrated in Fig. 19–4 for a zinc-titania glaze producing willemite (Zn_2SiO_4) crystals. The production of the crystals depends on heating the glaze above the nuclear forming temperature long enough to dissolve all but a few of the nuclei, then dropping at once to the growing temperature, which is held until any desired crystal size is arrived at. The shape of the crystals may be controlled by the growth temperature, and their color by added transition metal oxides.

Fitting the glaze to the body

Cause of crazing. As was shown in the case of glass, glazes are very weak in tension but quite strong in compression. Therefore, if the glaze has a higher coefficient of expansion than the body, on cooling it will go into tension and show the network of cracks known as crazing. The finer the network, in general, the greater the stress developed. On the other hand, very high compressive stresses may cause cracks known as peeling, a fault often occurring at edges and corners. Therefore, fitting a glaze to a body means more or less equalizing the coefficient of expansion of the two. However, the picture is complicated by the fact that the glaze as fixed in place may have a composition quite different from that computed from the formula, because of volatilization and solution of the body.

Measurement of glaze stresses. If a thin piece of body is glazed on one side only, it will curl under a stress developed in the glaze, and this amount of curl may be readily measured at any temperature in the following way. A tuning fork is made up of the body as shown in Fig. 19–5 and the outside of the prongs glazed. The whole is then

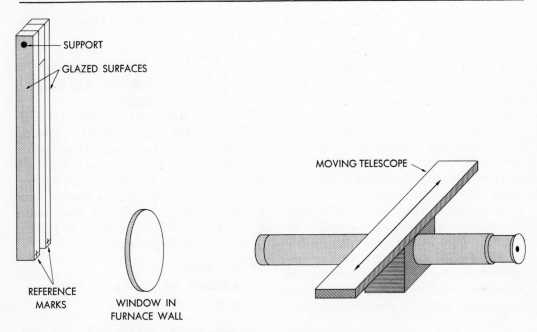

SUPPORT

GLAZED SURFACES

REFERENCE
MARKS

WINDOW IN
FURNACE WALL

MOVING TELESCOPE

Fig. 19-5. Method of measuring the stress in glazes.

fired to the normal maturing temperature and allowed to cool slowly in a furnace where the distance between the prongs may be measured with a telescope. If the prongs close up, the glaze is in compression and if they open up, it must be in tension.

Stresses in the glaze. Since glass, or the glaze, is strong in compression, but comparatively weak under tension, glazes with any great degree of tension craze in a network of cracks, the pattern of which becomes finer as the tension becomes greater. Therefore, the desirable glaze will always be in slight compression because its coefficient of expansion is smaller than that of the body. The stress in a normal glaze is shown in Fig. 19-6. It will be seen that above 550°C there is no stress, as the glaze is too soft to support it. The final stress at room temperature *a* is a considerable compression, but this decreases in a few days to point *b* because of the delayed contraction in the glaze.

Crazing is a very common defect in glazed ware, especially tiles. It can be controlled by careful design of glaze and body. There are many rules for preventing crazing by composition changes, but they do not always work. Only by a scientific approach through

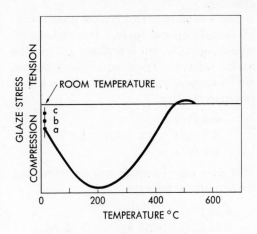

Fig. 19-6. Stress in a glaze.

stress measurements can this trouble be effectively controlled. It has often been noted that mat glazes are less subject to crazing than bright glazes. This is because the fine crystals act as a reinforcement to the glass and thus permit it to carry considerable tension without crazing.

Delayed crazing. Even a well-fitted glaze may show crazing after exposure to moisture for a considerable period. This is brought about by a partial rehydration of the body, which causes a slight expansion and consequent increased glaze tension. This effect is most pronounced with porous bodies, and is not found in the case of porcelain. Point *c* in Fig. 19–6 shows the stress after this test specimen had been exposed to moisture for some time. Had not the glaze been in high compression initially, crazing would have occurred.

The development of a good interface between the glaze and body is one of the best aids in preventing crazing. This is accomplished when the glaze and body are fired together.

Some examples of glazes

In this section are described a few typical glazes.

Raw lead glaze. This type of glaze is simple to make up and apply. It is rather soft, scratching easily, but is brilliant and may be readily colored. The composition given below by Binns is quite satisfactory for a bright transparent glaze.

$$0.6 \text{ PbO}$$
$$0.3 \text{ CaO} \qquad 0.2 \text{ Al}_2\text{O}_3 \qquad 1.6 \text{ SiO}_2$$
$$0.1 \text{ K}_2\text{O}$$

This glaze may be made mat by increasing the alumina as follows:

$$0.50 \text{ PbO}$$
$$0.35 \text{ CaO} \qquad 0.35 \text{ Al}_2\text{O}_3 \qquad 1.55 \text{ SiO}_2$$
$$0.15 \text{ K}_2\text{O}$$

To form an opaque enamel glaze, tin oxide is added:

$$0.72 \text{ PbO}$$
$$0.17 \text{ CaO} \qquad 0.17 \text{ Al}_2\text{O}_3 \qquad \begin{matrix} 1.93 \text{ SiO}_2 \\ 0.33 \text{ SnO}_2 \end{matrix}$$
$$0.11 \text{ ZnO}$$

All three of these glazes mature between 1050°C and 1120°C (cones 04 to 1). Details of preparation may be found in the appendix.

Bristol glazes. These are raw glazes used on terra cotta and stoneware where a lower maturing temperature than is possible with porcelain glazes is desired. A transparent glaze given by Wilson, maturing at 1175°C (cone 5), is quite satisfactory:

$$0.36 \text{ K}_2\text{O}$$
$$0.24 \text{ ZnO} \qquad 0.50 \text{ Al}_2\text{O}_3 \qquad 3.16 \text{ SiO}_2$$
$$0.40 \text{ CaO}$$

A mat, opaque, Bristol glaze, also from Wilson, is made by increasing the lime and zinc oxide thus:

$$0.24 \text{ K}_2\text{O}$$
$$0.27 \text{ ZnO} \qquad 0.39 \text{ Al}_2\text{O}_3 \qquad 2.00 \text{ SiO}_2$$
$$0.49 \text{ CaO}$$

Fritted glaze. These glazes are used for semivitreous and hotel chinaware. When properly made they have excellent working properties. Frits may be made in the laboratory by melting in clay crucibles, but it is more convenient to use commercial frits such as those listed by the companies making enamel frits.

A fritted glaze for a semivitreous body is given by Koenig as:

$$0.43 \text{ CaO}$$
$$0.26 \text{ PbO}$$
$$0.12 \text{ K}_2\text{O} \qquad 0.27 \text{ Al}_2\text{O}_3 \qquad \begin{matrix} 2.60 \text{ SiO}_2 \\ 0.31 \text{ B}_2\text{O}_3 \end{matrix}$$
$$0.06 \text{ Na}_2\text{O}$$
$$0.13 \text{ ZnO}$$

This matures at 1200 (cone 6).

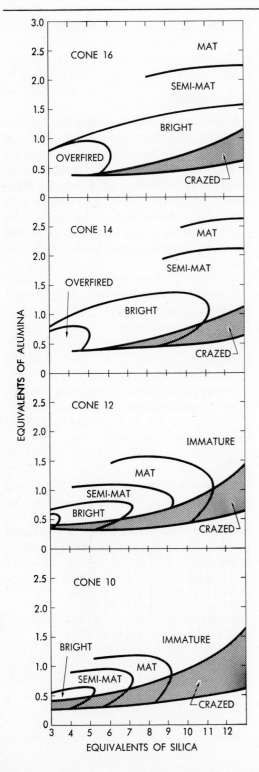

EQUIVALENTS OF ALUMINA

EQUIVALENTS OF SILICA

Porcelain glaze. The field of porcelain glazes has been carefully studied by Stull and by Howat and Sortwell. This type of glaze is easy to make up, nonpoisonous, hard, and has little tendency to craze. Only the high firing temperature prevents its greater use. In Fig. 19–7 are shown the composition fields of the porcelain glazes.

A bright glaze may be made up as follows to mature at 1250°C (cone 9):

$$\begin{matrix} 0.30 \ K_2O \\ 0.70 \ CaO \end{matrix} \quad 0.58 \ Al_2O_3 \quad 3.75 \ SiO_2$$

A mat glaze maturing at the same temperature is made by increasing the alumina:

$$\begin{matrix} 0.3 \ K_2O \\ 0.7 \ CaO \end{matrix} \quad 0.65 \ Al_2O_3 \quad 2.25 \ SiO_2$$

By using multiple alkaline earths, the maturing temperature may be reduced. A bright glaze maturing at 1190°C (cone 6) is given by Geller and Creamer.

$$\begin{matrix} 0.217 \ K_2O \\ 0.454 \ CaO \\ 0.135 \ BaO \\ 0.194 \ MgO \end{matrix} \quad 0.352 \ Al_2O_3 \quad \begin{matrix} 2.77 \ SiO_2 \\ 0.82 \ SnO_2 \end{matrix}$$

Crystalline glaze. Many types of glaze will produce crystals with the proper heat treatment as shown in Fig. 19–8. The following glaze produces excellent willemite crystals:

$$\begin{matrix} 0.235 \ K_2O \\ 0.087 \ CaO \\ 0.052 \ Na_2O \\ 0.051 \ BaO \\ 0.575 \ ZnO \end{matrix} \quad 0.162 \ Al_2O_3 \quad \begin{matrix} 1.700 \ SiO_2 \\ 0.202 \ TiO_2 \end{matrix}$$

Reduction glaze. These glazes contain copper or iron oxide to produce reds and greens respectively under the proper firing

FIG. 19–7. Compositions of the porcelain glazes ($RO = 0.3 \ K_2O, 0.7 \ CaO$).

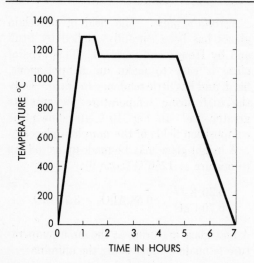

FIG. 19–8. Firing schedule to produce large crystals in a glaze.

conditions. These colors will be described in Chapter 22. It is also possible to produce these glazes by adding a reducing agent like ferrous iron or silicon carbide to the glaze or body and firing in an oxidizing atmosphere. The copper red glazes have been discussed by Mellor and Hetherington, who both conclude that they are of colloidal origin and contain copper in the metallic state. However, there is evidence that crystals of cupric oxide may account for the color in some cases.

Glaze defects

Crawling. The commonest defect encountered by beginners is crawling. This is caused by cracking of the drying layer of glaze, due to too high a shrinkage or poor adhesion to the body. On firing, the glaze usually draws back from the crack to give areas of unglazed biscuit. The cure is to be sure the biscuit is free from dust or oil and to use less raw clay, grind less, or add some gum.

Crazing. This has already been discussed.

Pinholes. This trouble may often be cured by longer or higher firing.

Uneven color distribution. This is due to specks of color and may be corrected by finer grinding of the stain with a portion of the glaze or by fritting the stain.

Orange peel surface. In spraying, the particles are not consolidated into an even layer. This condition can be improved by holding the spray gun closer to the work.

REFERENCES

ANDREWS, A. I., *Ceramic Tests and Calculations.* John Wiley and Sons, Inc., New York, 1928.

BINNS, C. F., *The Potter's Craft*, 3rd ed. D. Van Nostrand Co., New York, 1948.

BLAKELY, A. M., "Life History of a Glaze: I, Maturing of a Whiteware Glaze." *J. Am. Ceram. Soc.* **21**, 239, 1938.

BLAKELY, A. M., "Life History of a Glaze: II, Measurement of Stress in a Cooling Glaze." *J. Am. Ceram. Soc.* **21**, 243, 1938.

FOSTER, H. D., *Salt Glazes on Structural Clay Building Units.* Ohio State University, Exp. Sta. Bull. No. 113, 1931.

KOENIG, J. H., and EARHART, W. H., *Literature Abstracts of Ceramic Glazes.* College Offset Press, Philadelphia, 1951.

MELLOR, J. W., "The Chemistry of the Chinese Copper-red Glazes." *Trans. Ceram. Soc.* **35**, 364, 1936; **35**, 487, 1936.

NORTON, F. H., "The Control of Crystalline Glazes." *J. Am. Ceram. Soc.* **20**, 217, 1937.

PARMELEE, C. W., *Ceramic Glazes.* Industrial Publications, Inc., Chicago, 1948.

SEGAR, H. A., *Collected Writings.* Chemical Publishing Co., New York, 1902.

SORTWELL, H. H., "High Fire Porcelain Glazes." *J. Am. Chem. Soc.* **4**, 719, 1921.

STULL, R. T., "Influences of Variable Silica and Alumina on Porcelain Glazes of Constant R O." *Trans. Am. Ceram. Soc.* **14**, 62, 1912.

CHAPTER 20

ENAMELS ON METAL

Introduction

Enamels form an excellent protective coating for metals. They are durable and washable, and may be made in white or various colors. The enamel ware industry is primarily a metallurgical one, and buys the enamel in many cases from manufacturers who also supply the directions for application and firing. The industry is divided into three divisions; sheet steel enameling, cast iron enameling, and specialties such as signs and jewelry.

Low fusing glasses

As the base metal on which the enamel is placed oxidizes and warps if heated to a high temperature, enamels must have a low softening temperature. Therefore, it is of interest to study the means of forming low melting glasses.

The silica-oxygen network is a very stable one, but partial replacement of the silica with boron forms a portion of boron-oxygen triangles much weaker in bond strength than the tetrahedrons. Thus boron is an almost universal constituent of enamels. It is also possible to replace part of the oxygen by the weaker bonded fluorine in the glass network and thus give a lower softening glass; therefore, it will be found that many enamels contain fluorine.

The more active network modifiers produce a lower melting glass. For example, lithium may be used to replace sodium or potassium. The polarizable ions such as Pb^{++} weaken the network bonds as increas-

ing amounts are added, and for this reason some enamels contain this element. Lead makes an easy flowing, brilliant enamel, but for reasons of health it is limited to cast iron enamels. Today, few lead-containing enamels are used.

At the same time that the bonds are weakened to provide greater fluidity, there is a tendency to develop less chemical resistance. However, it is possible to obtain enamels with sufficient resistance by a careful selection and proportioning of the ingredients.

Enamel frits. Unlike most glazes, enamel contains almost 100 per cent of frit. The frit is melted in rotary furnaces or small glass tanks, quenched in water and crushed.

Milling. The frit, together with mill additions such as clay and the opacifier, is wet ground in a ball mill to a definite fineness, perhaps 3 per cent on a 200-mesh screen. The enamel slip is often aged 24 hours and is then ready for application.

Theories of adherence

One of the problems of enameling is to provide a good adherence between the enamel glass and the metal. A great amount of study has been given this problem, and yet there is no generally accepted theory as to the mechanism of adherence. It has been shown that a small amount of one of the transition elements is required. Some of the theories for adherence of enamel to steel are:

1. The transition element, preferably cobalt, promotes the growth of dendritic

iron crystals from the base into the enamel. The crystals thus serve as anchors.

2. Cobalt causes an adherent oxide coat on the iron base, into which the enamel fuses.

3. The transition elements are polarizable and therefore promote chemical bonds between the metal and glass.

Methods of obtaining opacity

In most cases it is desirable to produce a white or light-colored enamel. As the base coat is dark-colored because of the transition elements, it is usually necessary to cover this with an opaque coat. A high degree of opacity is desired, since the enamel thickness must be low for mechanical and cost reasons. Therefore, much work has been done in this field.

Theory of opacifiers. The opacity is obtained by distributing in the enamel glass small particles having a refractive index different from that of this glass. The scattering power increases as the difference in index between the two increases, and as the particle size approaches the wavelength of light.

The index of refraction of the enamel glass cannot be greatly varied and remains at about 1.5. However, particles of higher or lower index of refraction are available as listed in Table 20–1.

Opacifiers added to the enamel. One of the usual methods of producing opacity is to add the finely divided particles to the batch in the mill. During the short time the enamel is heated, there is little chance for solution. Formerly SnO_2 was used for this purpose, but now the less expensive zirconium and titanium compounds have largely displaced it.

Opacifiers developed in the enamel glass. As was shown for the crystalline glazes,

Table 20-1

Properties of Opacifiers

Material	Index of refraction	Theoretical relative scattering power
TiO_2	2.7	1.2
ZrO_2	2.2	0.7
Sb_2O_3	2.1	0.6
ZnO	2.0	0.5
SnO_2	2.0	0.5
$ZnAl_2O_4$	1.9	0.4
Al_2O_5	1.8	0.3
$MgAl_2O_4$	1.7	0.2
Enamel glass	1.5	0
CaF_2	1.4	0.1
NaF	1.3	0.2
Air	1.0	0.5

crystals may be grown in the glass at the proper temperature. While this procedure is possible in the case of enamels, the short firing time precludes any extensive development, and this method is not useful.

Jewelry enamels

Base metals. These enamels are generally used on copper or copper alloys, but gold and silver are quite suitable. As these enamels are generally applied in color patterns, some means must be provided to keep one color from diffusing into the next. Three methods are used for this purpose. One, called *cloisonné*, outlines the pattern by fences of metal ribbon, hard soldered to the base as shown in (a) of Fig. 20–1. The enamel slip is then run into each separate compartment and fired (b). After cooling, the whole surface is polished down as shown in (c), leaving the partitions showing as fine metal lines between the colors.

In the second method, called *champlevé*, the dividing lines are applied in wax and

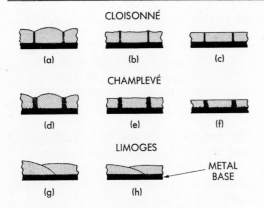

CLOISONNÉ

(a)   (b)   (c)

CHAMPLEVÉ

(d)   (e)   (f)

LIMOGES

METAL BASE

(g)   (h)

FIG. 20–1. Section through enameled jewelry to show the three common types of application.

then etched to take out the metal between, which is later replaced by enamel as shown in (d), (e), and (f) of Fig. 20–1. In some cases the metal is cut out with an engraving tool.

The third method, often known as *Limoges*, applies the enamel directly without any partitions, as shown in (g) and (h) of Fig. 20–1, so that the colors diffuse into each other to some extent. Often several firings are employed before the piece is finished.

Composition of jewelry enamels. As these enamels must be fired at a lower temperature than those used for iron, they contain numerous fluxes to produce a low melting glass. A typical formula for an enamel to be used on copper is:

SiO_2	20
Red lead	61.5
Sodium nitrate	13.7
Boric acid	26.6

Frit, grind and pass through a 100-mesh screen. Fire at 950°C for 2 minutes.

Enamels for sheet steel

Base metal. Enameling iron must have a low content of impurities in order to produce a good coating. The most generally used metal is Armco iron with the following analysis:

C	0.015 per cent
Mn	0.020
P	0.005
S	0.025
Si	trace

This sheet metal is shaped by pressing or drawing and must then be thoroughly cleaned before applying the enamel.

The ground coat. The ground coat is then applied by spraying to a weight of about two ounces per sq ft. A typical ground coat frit batch is:

Feldspar	30	per cent
Flint	20	
Borax	30	
Soda ash	8	
Fluorspar	6	
Sodium nitrate	4	
Cobalt oxide	0.5	
Nickel oxide	0.3	
Manganese dioxide	1.2	

The sodium nitrate is an oxidizing agent and the transition elements are added to give adherence to the iron. This batch is melted in a frit furnace, and quenched in water. It is then mixed with the following additions and put in a ball mill.

Frit	100
Clay	7
Borax	0.5
Magnesium carbonate	0.12
Water	45

The clay, a special fine grained material, is used for a suspending medium, but it also contributes to the opacity. The borax and magnesium carbonate are electrolytes to aid suspension. The grinding is carried on until 74 per cent remains on a 200-mesh screen. After application the layer is dried rapidly and then fired in a muffle furnace for about $2\frac{1}{2}$ minutes at 830 to 885°C, the higher temperatures applying to the heavier gauge metals.

Cover coat. A typical cover coat frit batch is:

Feldspar	25.0 per cent
Borax	25.5
Flint	21.0
Soda ash	3.0
Sodium nitrate	3.5
Cryolite	14.0
Zinc oxide	5.0
Antimony trioxide	3.0

This is fritted as before and the mill batch is:

Frit	100.0
Clay	7.0
Tin oxide	8.0
Borax	0.5
Water	50.0

This is ground, with 3 to 6 per cent remaining on a 200-mesh screen. It is sprayed on the ground coat at a weight of about 6 ounces per sq ft. The firing is for about 3 minutes at 800 to 850°C. For high grade work two cover coats are used, the second one fired about 10°C lower than the first.

A considerable amount of enamel is now made without a ground coat by treating the iron surface with a soluble nickel salt and applying the cover coat directly. This cuts the enameling cost but does not give as good mechanical properties.

A typical flow sheet for an enameling plant is shown in Fig. 20–2.

Cast iron enamel

Base metal. The cast iron is the normal type of gray iron with a rather high silicon content to give fluidity and to prevent chill-ing in thin sections. A typical analysis is:

Carbon	3.25–3.60
Silicon	2.25–3.00
Manganese	0.45–0.65
Phosphorus	0.60–0.75
Sulphur	0.05–0.10

The castings are cleaned up by grinding and sand blasting to give a bright surface.

Ground coat. A thin ground coat of some-what the same composition as that used on sheet steel is then sprayed on at once.

Cover coat. While some of the small cast-ings are coated by spraying as for sheet steel, the cover coat for larger pieces is generally applied as a dry powder to the heated iron. This process seems cumbersome, but pro-duces excellent enamel surfaces. Since there is a large percentage of reclaimed powder to be remelted, the cast iron enameler cannot economically buy the frit, but must smelt it himself. A typical composition is:

Feldspar	8.7 per cent
Borax	11.4
Quartz	25.5
Soda ash	14.1
Sodium nitrate	5.1
Red lead	7.8
Zinc oxide	5.1
Whiting	3.0
Antimony trioxide	10.8
Titanium dioxide	6.8
Sodium silicofluoride	1.7

The frit is dry ground and is shaken onto the piece after being preheated in a muffle furnace. The hot piece is handled by lifting forks and then placed on a manipulator so that it can be turned as the enamel is sifted onto it. This operation requires great skill and because of the radiant heat is a rather uncomfortable task. The manufacturers have been very backward in not developing

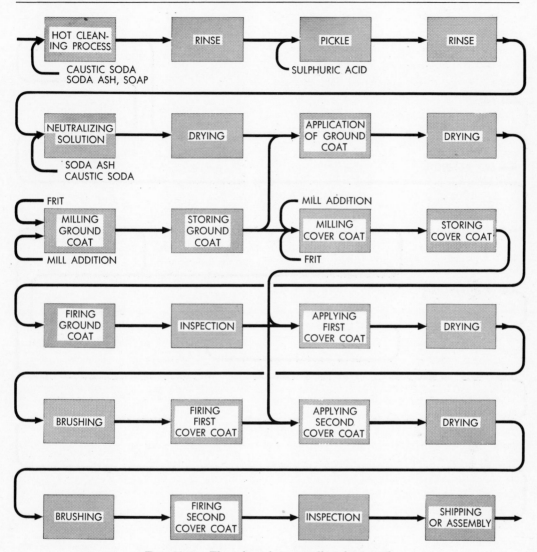

FIG. 20–2. Flow sheet for enameling sheet steel.

better handling equipment and automatic enameling machines. About one-third of the frit used falls on the floor and must be reclaimed. Because it picks up scale and other impurities, it cannot be used again, but must be refritted. It is probable that methods could be developed to process this frit for direct re-use.

As the enamel powder falls on the hot iron, it fuses and sticks in place. After about a $\frac{1}{16}$ inch has been built up, the piece is heated again, and a second coat put on which gives a total thickness of $\frac{3}{32}$ to $\frac{1}{8}$ inch. A flow sheet for this process is shown in Fig. 20–3.

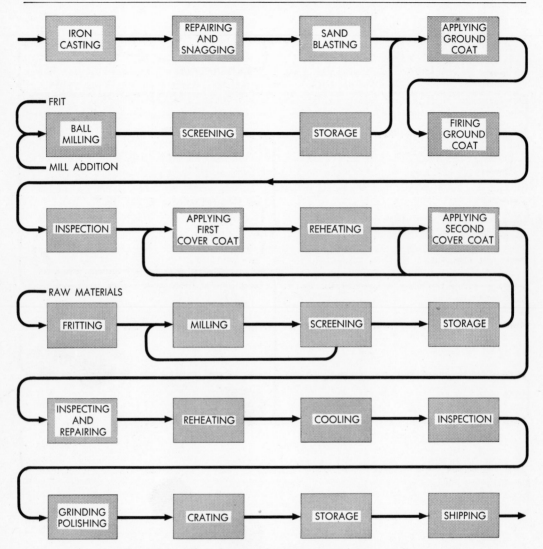

FIG. 20–3. Flow sheet for dry process cast iron enameling.

Acid resisting enamels. The usual cover coat is readily stained by fruit juices, so an increasing proportion of enamels are now made more stable by an increase in the silica content and a change in the fluxes. Lithia has been found to be very useful in these enamels.

Enamel defects. The enameler is sometimes cursed with defects that cause rejection of the piece. Often it is difficult to find the cause, since no obvious change in operation synchronizes with this trouble. As these defects are fully described in the references, they will not be discussed here.

REFERENCES

ANDREWS, A. I., *Enamels*, 2nd ed. The Twin City Printing Co., Champaign, Ill., 1945.

CUNYNGHAME, H. H., *On the Theory and Practice of Art Enamelling on Metals*. Archibald Constable and Co., Westminster, England, 1899.

McCLELLAN, E. H., "Enamel Bibliog. and Abstracts." *J. Am. Ceram. Soc.*, Columbus, Ohio, 1944.

STALEY, H. F., *Materials and Methods Used in the Manufacture of Enameled Cast Iron Wares*. Natl. Bur. Stds., Tech. Paper 142, 1919.

TURK, KARL, *The Ready Remedier*. The Baltimore Porcelain Enameling and Manufacturing Co., Baltimore, 1931.

WEYL, W. A., "Adhesion to Glass." *Am. Soc. Test. Mat.* **46**, 1500, 1946.

WOLFRAM, H. G., and HARRISON, W. N., "The Development of Some Jewelry Enamels." *J. Am. Ceram. Soc.* **7**, 857, 1924.

CHAPTER 21

COLOR FORMATION IN GLASSES AND GLAZES

Introduction

One of the most fascinating subjects in the field of ceramics is the production of colors. Ceramic colors are of great variety and, compared to other colors, are remarkably permanent. In this chapter the mechanism of color formation will be discussed and the more important coloring elements described.

Mechanism of color formation in glasses

Properties of light. Visible light is a very narrow region in the long spectrum of electromagnetic waves that travel in the ether. The wavelength of the longest visible red waves is about 700 millimicrons and of the shortest visible violet about 400 millimicrons.

The normal human eye is a marvelous instrument evolved through millions of years from a most rudimentary beginning in the skin of some aquatic animal. Color vision is not by any means the rule in the animal kingdom. Only man, the higher apes, birds, lizards, turtles, and fish are known to have it.

While the color receptors of the human retina are not completely understood, it is known that there are responses to three regions of the spectrum, which combined give the effect of color. If either one or two of these response mechanisms are lacking, color blindness results. However, the sensitivity of the normal eye is not equal over the whole length of the spectrum. It reaches a maximum in the yellow-green region, as shown by the curve in Fig. 21–1.

Definition and measurement of color. Color, to a physicist, is a vibration in the ether; to a physiologist it is a stimulus to the retina; and to a chemist it is dye. The ceramist looks at color in a relative sense in that he compares his product with some natural color, as the terms lilac, oxblood, or peach blow indicate.

The spectrum colors are defined by well-known terms and cover a wavelength range as follows:

Red	700–620 millimicrons
Orange	620–592
Yellow	592–578
Green	578–500
Blue	500–450
Violet	450–400

All colors except metallic lusters are the result of selective absorption of light transmitted through a transparent or translucent medium. A yellow glaze looks colored because light falling on it passes into the glaze layer and is reflected out again; in the process some of the blue and the red are absorbed, leaving the yellow to predominate in the emergent beam. It is well known that the apparent color of an object depends on the kind of light that illuminates it. A white paper looks red in red light, and the shadows on newly fallen snow look blue, for they are illuminated by the relatively weak, but bluish light from the sky. It is a common experience to have two colors match well in daylight and to find later that they are quite different in artificial light.

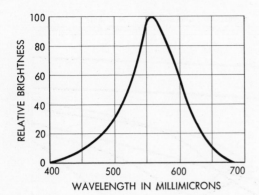

FIG. 21-1. Color sensitivity of the human eye.

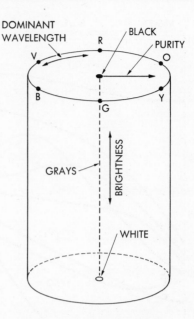

FIG. 21-2. Color cylinder.

There are two general methods of measuring color. One is a comparison method with a series of standard samples, for example, those supplied by the Munsell system. Here the different colors are arranged in the form of a color cylinder, as shown in Fig. 21-2. The hues are distributed around the circumference with white, grays, and black on the axis. As this axis is approached, the colors become more gray or have less chroma. The colors lower on the cylinder are light or brilliant and the higher they go the darker they become. Each color has its complement directly opposite on the cylinder. Two colors are said to be complementary when they give a neutral shade on blending. This blending may be done on the familiar spinning disk, or by actual mixing. The color cylinder has around 800 numbered colors, so that a comparison may be readily made with a specific ceramic specimen and designated by number.

As the color samples are not completely permanent, a more scientific method of measurement may be desirable. This can readily be done by means of the spectrophotometer, which records the transmission or reflection of each wavelength in the spectrum. In Fig. 21-3 are shown color curves of several glazes. The curves may be converted into three color specifications expressed as dominant wavelength (hue), saturation, or the amount of white light mixed with monochromatic light of the dominant wavelength (chroma), and intensity (brilliance). For a more detailed discussion of color measurement the student should consult Hardy.

It should be kept clearly in mind that we have no method of measuring some of the more intangible factors connected with color. This is particularly true in ceramics, where translucency or transparency give a depth to color that is a large part of its charm. It is quite possible to find a Munsell color sample of ink on paper to match the color of a thick celadon glaze on an old Chinese vase, but the glaze has a depth that is quite absent from the ink.

Selective absorption by ions. Everyone is familiar with the fact that certain salts dis-

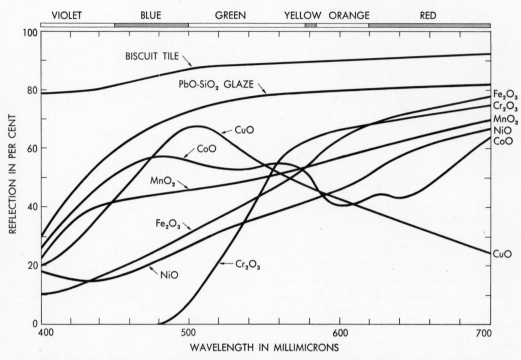

FIG. 21-3. Color curves for a simple lead **glaze** with various transition ions in solution.

solved in water produce a colored solution. This is caused by selective light absorption by one of the ions. Ions are known to absorb light energy in three ways; one, by the vibration of the atom as a whole, absorbing in the infrared region; two, by vibration of the electrons, absorbing in the ultraviolet region; and three, by orbit jumps, absorbing in the visible region. It is this last type of energy absorption that interests us here.

Coloring elements. Not all ions have the electron configuration that permits absorption in the visible range. Only those elements with an incomplete electron shell, such as the transition elements and the rare earth elements, are capable of giving ionic absorption.

Color modifiers. A colored ion does not always give the same color, for the rate of electron vibration is influenced by the environment. For example, the valence state, the position in the glass network, and the type of ions surrounding it all influence the color of an ion.

Chromophore colors. There are certain cases where a color absorbing complex is formed, much like organic dye stuffs, for example in compounds of antimony and cadmium. These complexes are not stable at the higher temperatures, but often give brilliant colors in enamels, low temperature glazes, or overglaze decorations.

Solution colors

Ions of the transition elements. The ions of the transition elements are shown in Table 21-1 with their probable colors, both when in the glass network and when in the modifier position. As will be seen, our information is not complete by any means,

Table 21-1

Colored Ions in Glasses

Ion	In glass network		In modifier position	
	Coordination number	Color	Coordination number	Color
Cr^{2+}		---		blue
Cr^{3+}		---	6	green
Cr^{6+}	4	yellow		---
Cu^{2+}	4	---	6	blue-green
Cu^{+}		---	8	colorless
Co^{2+}	4	blue-purple	6-8	pink
Ni^{2+}		purple	6-8	yellow-green
Mn^{2+}		colorless	8	weak orange
Mn^{3+}		purple	6	---
Fe^{2+}		---	6-8	blue-green
Fe^{3+}		deep brown	6	weak yellow
U^{6+}		orange	6-10	weak yellow
V^{3+}		---	6	green
V^{4+}		---	6	blue
V^{5+}	4	colorless		---

but a general picture may still be drawn.

A more definite picture may be found in the case of a specific glaze, $PbO \cdot SiO_2$, into which have been introduced the transition ions. The results are shown in Fig. 21–3 by means of spectrophotometric reflection curves and in Table 21–2 by color names. The more important coloring ions are seen to be Cu^{++}, Cr^{6+}, Mn^{+++}, Fe^{+++}, Co^{++}, Ni^{++}, and U^{6+}.

There are a few special cases that might be mentioned, such as the pure blue obtained from Cu^{++} in an alkaline glaze, and the pink from Co^{++} in phosphate glasses.

Rare earth ions. These ions differ from the transition ions in that the electron vibration causing absorption is not in the outer shells but in a more protected position inside. For this reason the vibration rate is not influenced by adjoining atoms, and the ab-

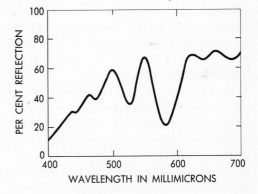

FIG. 21–4. Reflection curve of a neodymium-containing glaze.

sorption spectrum is banded, rather than continuous as it is for the transition ions.

A rare earth glaze containing Nd^{+++} ions gives a reflection spectrum as shown in Fig. 21–4. The rare earths do not give brilliant

Ion	Color
Ag^{++}	pale green
Au^{++}	faint violet[1]
Ca^{++}	intense blue
Cr^{6+}	yellow-orange
Cu^{++}	intense green
Fe^{+++}	yellow
Mn^{+++}	purple
Mo	faint green
Ni^{++}	yellow-green
Pd^{++}	gray-black[1]
Pt^{++}	gray-black[1]
Ti^{++}	faint yellow
V^{4+}	faint yellow
U^{6+}	intense yellow-red
W^{6+}	pale yellow

[1] May be colloidal colors.

colors, but are used in filters because of their sharp cut-off and in some art glasses for their soft, dichromatic effects.

Colloidal colors

Mechanism of colloidal color formation. A series of particles dispersed in a transparent medium with a different index of refraction scatters the light passing through, as explained in Chapter 20. However, in colloid colors we are concerned with much smaller particles. They are so small compared with the wavelength of light that they do not prevent complete transparency. Colloidal colors are most commonly produced by gold, silver, and copper dispersed in the metallic state as colloidal particles on the order of 50 millimicrons in diameter. The particles themselves have a selective absorption and a complementary selective reflection.

As the particle size changes, there is a change in position of the absorption bond, but as yet the complete mechanism of colloidal color is not understood.

Gold colloid colors. Gold is dissolved in glass, not as the metal, as some have believed, but rather in the oxidized state as Au^+ ions which fit into the glass network in the same manner as K^+ or Na^+.

In some glass compositions, the Au^+ ion is stable enough to be frozen in on quick cooling so that the quenched glass is colorless. In other cases, especially with reducing agents, the Au^+ reduces to the metal even with quick cooling, and color is produced at once. Most gold ruby glasses, however, are formed by reheating the colorless, quenched glass for a sufficient period to decompose the Au^+ into metallic gold, thus:

$$3Au^+ = Au^{+++} + 2Au.$$

However, most gold rubies contain a small amount of SnO_2 or other variable valence oxide to decrease the solubility of gold by a reaction such as

$$2Au^+ + Sn^{++} = Sn^{++++} + 2Au.$$

The size of the metallic gold particles depends on the time and temperature of heat treatment, which in turn influences the hue. It is estimated that the gold content is about .0001 gm per cc and the size is as follows:

Pink	4–10 millimicrons
Ruby	10–75
Blue	75–110
Livery brown	110–170
Diffusion without color	400–700

In Fig. 21–5 is shown a section of a gold ruby greatly magnified to indicate the scale of the colloidal gold particles. The best gold rubies seem to occur in lead glasses. The composition $K_2O \cdot PbO \cdot 6SiO_2$ plus 0.0075

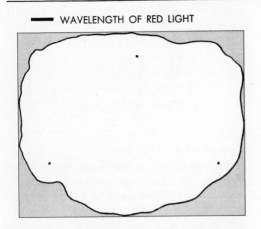

FIG. 21–5. Enlarged section of a gold ruby glass (7000 times).

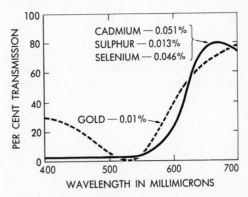

FIG. 21–6. Transmission curves of two types of red glass.

per cent of gold gives good results when reheated to 500–700°C. A composition of $Na_2O \cdot CaO \cdot 6SiO_2$, plus 1 per cent of Al_2O_3 and 0.0075 per cent of gold, if reheated to 650°C produces a good red. About 10 per cent of the Na_2O should be derived from sodium nitrate to give an oxidizing condition in the melt. The addition of $\frac{1}{2}$ per cent of SnO_2 also helps to form a good red.

The transmission curve of a gold ruby is shown in Fig. 21–6. The transmission in the blue end gives the characteristic purplish red to this glass.

Copper colloid colors. The copper colloids are similar to those of gold. Cu^+ ions are in the network of the molten glass, but with a mild reduction on cooling form metallic copper as follows:

$$2Cu^+ = Cu^{++} + Cu,$$

or $$2Cu^+ + Sn^{++} = Sn^{++++} + 2Cu.$$

The amount of copper needed is much greater than for gold, 0.1 to 0.5 per cent. Therefore, the copper, if in the cupric form, will give a green. There is evidence that some of the red copper colors may be due to the red copper oxide, CuO, dispersed as

crystals throughout the glass. The whole subject of reduced copper colors needs more study.

Silver colloid colors. The silver ion Ag^+ is more stable than that of gold, so that 0.2 to 0.5 per cent of this metal is required to cause precipitation. If reducing agents are present, a lower amount will suffice. The characteristic color of the silver colloid is yellow.

Carbon and sulphur colloids. Both of these elements are believed to form colloids in glasses, giving ambers and browns.

Cadmium-selenium colors. The important red glass used for signal lights is believed to be a colloidal dispersion of particles made up of a solid solution of CdS and CdSe. A transmission curve of this glass shown in Fig. 21–6 indicates the almost complete lack of transmission in the blue end of the spectrum.

Colors in crystals

Some glasses and glazes are colored by means of crystals dispersed throughout the mass. An important example is the scarlet glass made by the early Egyptians which is colored by crystals of red copper oxide. Another common color is due to the bright red crystals of Pb_2CrO_6. The latter are unstable above 1000°C and transform to green CrO_2 crystals.

Many other crystals may be colored by taking into solid solution one of the transition elements. For example, the colorless willemite crystals Zn_2SiO_4 may be colored as follows:

Cu — light green
Fe — gray
Mn — yellow
Cr — gray
Co — intense blue

REFERENCES

BADGER, A. E., WEYL, W., and RUDNOW, H., "Effect of Heat Treatment on Color of Gold Ruby Glass." *Glass Industry* **20**, 407, 1939.

HAINBACK, R., *Pottery Decorating*. Scott, Greenwood and Son, London, 1924.

HARDY, A. C., *Handbook of Colorimetry*. Technology Press, Cambridge, Mass., 1936.

NICKERSON, D., "Color and Its Description." *Bull. Am. Ceram. Soc.* **27**, 47, 1948.

PARMELEE, C. W., *Ceramic Glazes*. Industrial Publications, Inc., Chicago, 1948.

SHAW, D. T. H., "Color Formation in Raw Lead Glazes." *J. Am. Ceram. Soc.* **15**, 37, 1932.

STOOKEY, S. D., "Coloration of Glass by Gold, Silver and Copper." *J. Am. Ceram. Soc.* **32**, 246, 1949.

WALLS, G. L., *The Vertebrate Eye*. Cranbrook Institute, Bloofield Hills, Michigan, 1942.

WEYL, W. A., "Coloured Glasses." *J. Soc. Glass Technol.* I, 133, 1943; II, 265, 1943; II, 158, 1944; III, 267, 1944; IV, 289, 1945; V, 90, 1946.

CHAPTER 22

CERAMIC STAINS AND COLORS

Introduction

In the preceding chapter a general discussion of color formation was given. This chapter will attempt to give some data on colored crystals used as stains, either in the glaze, over the glaze, or under the glaze.

Colored spinels

A stain must be made from colored crystals of very limited solubility in the glaze. For this reason the very stable spinels are most used by the color manufacturers. The typical spinel structure, represented by $B^{+2}A_2^{+3}O_4$, has a unit cell of 32 oxygen atoms, 8 B atoms in tetrahedral positions, and 16 A atoms in octahedral positions, as shown in Fig. 22–1. There is, however, a second type of spinel, $A^{+3}B^{+2}A^{+3}O_4$, in which the unit cell has 32 oxygen atoms, 8 A atoms in the tetrahedral position, 8 B atoms in the octahedral position and 8 A atoms in the octahedral position. The spinel lattice is very flexible and permits solid solution by substitution for B^{+2} or A^{+3} of other atoms as long as their ionic radii do not differ too much. Spinels are so flexible that differences greater than the usual maximum difference of 15 per cent can be tolerated.

Colored spinels may be formed with the B and A atoms shown in Table 22–1. The barium, calcium, and strontium compounds are probably not true spinels, but nevertheless are valuable colored crystals and will be included with the spinels.

The spinels are prepared by grinding together the two oxides with a slight excess of the basic one over the stoichiometric proportions. Also it is usually necessary to add 30 per cent of B_2O_3 to accelerate the reaction. The batch is calcined from 900 to 1300°C for 24 hours. It is then ground, leached with 10 per cent HCl to remove any remaining B_2O_3, washed, and ground to 1 to 5μ in size. Size control is very important in ceramic colors because brilliance is lost with coarse particles, and too great a solution in the glaze occurs if they are too fine. The importance of thorough washing to remove soluble salts cannot be too greatly emphasized.

Some of the colored spinels are listed in Table 22–2. It will be seen that some of them are unstable and revert to the solution colors of the cations, while others hold their original hue. This is indicated by S (stable) or PS (partially stable) following the color description. On the whole the aluminates are the most stable, the chromates next, and the ferrites least stable.

An interesting example of color production in spinels is shown in the $MgO \cdot Al_2O_3 \cdot Cr_2O_3$ system. As Cr_2O_3 gradually replaces Al_2O_3, shown in Fig. 22–2, the lattice is expanded. At the low Cr_2O_3 end a red crystal results, stable to the highest porcelain temperatures, while at the high Cr_2O_3 end the predominant form is green. Between 30 and 45 mol per cent of Cr_2O_3, both forms are stable.

Other colored stains

Silicates. Cobalt silicate made by calcination of flint and cobalt oxide, sometimes with a little flux such as feldspar, is a com-

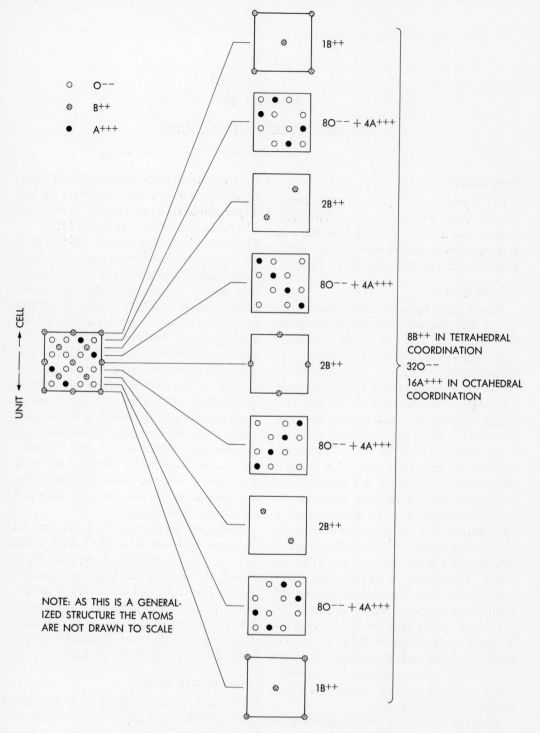

FIG. 22–1. A unit cell of the normal spinel structure.

Table 22-1

Cations in Colored Spinels

B^{+2}	Ionic radius	A^{+3}	Ionic radius
Co^{+2}	0.82	Al^{+3}	0.57
Cu^{+2}	0.70	Fe^{+3}	0.67
Fe^{+2}	0.83	Cr^{+3}	0.64
Ni^{+2}	0.78	Mn^{+3}	0.70
Mg^{+2}	0.78	Co^{+3}	0.72
Mn^{+2}	0.91		
Zn^{+2}	0.83		
Ba^{+2}	1.43		
Ca^{+2}	1.06		
Sr^{+2}	1.27		

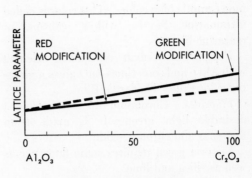

FIG. 22–2. Colored spinels in the $MgO \cdot Al_2O_3 \cdot Cr_2O_3$ system.

Table 22-2

Colors Produced by Spinels

Spinel	Crystal color	Under porcelain glaze	Under lead glaze	In a Parian body
$CoAl_2O_4$	deep blue	deep blue, PS	brilliant blue	blue, D
$CuAl_2O_4$	apple green	gray-green	green	gray-green
$MnAl_2O_4$	tan	tan, S	tan	brown, S
$NiAl_2O_4$	sky blue	green	yellow-green	gray-green
$BaCr_2O_4$	dark green	sage green, S	sage green, S	green, S
$CuCr_2O_4$	dark green	sage green, S	sage green, S	green, S
$CoCr_2O_4$	deep blue-green	blue-green, S	blue-green	blue-green, S
$MgCr_2O_4$	dark olive green	green	scummed	yellow-green, S
$MnCr_2O_4$	dark brown	yellow-green, S	scummed, S	tan, S
$NiCr_2O_4$	leaf green	gray-green	mottled	dark green
$SrCr_2O_4$	dark green	green, S	sage green, PS	bright green, S
$ZnCr_2O_4$	gray-green	brown	mottled, S	light brown
$BaFe_2O_4$	medium gray	gray	brown	brown
$CaFe_2O_4$	medium gray	gray	tan	tan
$CoFe_2O_4$	black	gray	gray, S	gray, S
$CuFe_2O_4$	dark brown	gray	gray	tan
$MgFe_2O_4$	orange brown	brown, PS	tan, PS	brown, S
$MnFe_2O_4$	dark gray	brown	brown	brown, S
$NiFe_2O_4$	black	gray	gray	brown, PS
$SrFe_2O_4$	medium gray	gray	tan	brown, S
$ZnFe_2O_4$	dark gray	gray	orange-brown	brown

mon ceramic blue color. This color is called ultramarine, Sèvres, willow, canton, or mazarine blue.

Phosphates. Cobalt phosphate calcined with Al_2O_3 and sometimes ZnO gives a pleasing color called Thénard's Blue.

Fluorides. Chromium fluoride makes a desirable light green. It is made from fluorspar and chromic oxide.

Victoria green requires some fluorspar as well as silica and lime.

Antimonates. Naples yellow, much used in ceramics, is largely lead antimonate, but also contains some lime, alumina, and tin oxide for stabilization.

Uranates. Calcium and sodium uranate form the basis for some of the most brilliant red, orange, and yellow colors, but present restrictions make their use in ceramics impossible.

Other crystalline colors. Cobalt oxide and magnesia form a color called Berzelius' Pink, which is quite stable. In this case the Co^{2+} must be in 6- or 8-fold coordination.

Stains colored by colloids. The chrome-tin pinks and reds are much used as fairly stable stains. Apparently the colloidal chromium particles are stabilized in crystals of calcium stannate. A typical formula for such a stain is:

Stannic oxide	50 per cent
Calcium carbonate	25
Potter's flint	18
Borax	4
Potassium bichromate	3
Calcine to 1250°C	

A still more stable red stain is made by stabilizing the chromium colloid in alumina, as found in nature in the ruby. The mixture

Aluminum hydroxide	85 per cent
Chromic oxide	6
Boric acid	9

calcined to 1500°C gives an excellent pink.

Gold colloid stains are very important in pink, purple, and carmine enamel colors as well as for underglaze use on high fired porcelain. A gold solution is reduced to precipitate the colloid onto the surface of a stable material like kaolin. The mass is then calcined to give the pink or red.

Overglaze colors

Overglaze colors are essentially a mixture of a stain, an opacifying agent, a flux, and a diluent. On firing at a low temperature, only a little above red heat, the flux melts and combines with the highly viscous surface of the glaze. In some cases the overglaze color only adheres to the glaze surface, and in others it penetrates quite deeply into the glaze itself. Overglaze colors are usually made opaque by the addition of SnO_2, ZnO, or other opacifying agent. In other words, they are a low fusing enamel type glaze.

All of the colors used on one piece should mature at the same temperature, for it is not economical to make several overglaze fires. The fired color should be reasonably resistant to wear and chemical action of soaps and food acids.

The fluxes used with overglaze colors are generally lead borosilicates. Hainback gives some fluxes as shown below:

PbO	$1.00\ SiO_2$	For general use
	$.70\ B_2O_3$	
$.25\ PbO$	$2.95\ SiO_2$	For gold-reds
$.75\ Na_2O$	$1.50\ B_2O_3$	
PbO	$3.40\ B_2O_3$	For pale colors

The diluents used in overglaze colors may be raw or calcined kaolin, silica, or alumina. These are used only in the weaker colors. The vehicle used for applying overglaze colors may be a solution of water soluble gum, such as gum tragacanth or, more commonly, an oil thinned with turpentine. Methods of applying will be discussed in the next chapter.

Firing overglaze colors. The overglaze colors must be fired under conditions that protect them from combustion gases, especially those containing sulphur. These colors have usually been fired in a muffle kiln, hence the name "muffle colors" is often applied to them. In modern potteries the overglaze decoration is fired in small continuous tunnel kilns, often electrically heated. The temperature is around 750°C (cones 015–017).

Underglaze colors

Colors under the glaze are thoroughly protected and they withstand wear much better than overglaze colors. For this reason hotel china is always underglazed. On the other hand, the color palette is more limited because of the higher firing temperature. These colors are made from stains, diluents and often fluxes.

Fluxes. The fluxes used in underglaze colors are less in amount and of higher softening point than for overglaze colors. Fluxes often used are feldspar, and glaze frit. They are added in the amount of 20 to 30 per cent and need not be as powerful as in the overglaze colors, because of the higher temperature of firing and the fact that they need not melt down to a glass.

Diluents. These materials are inert and merely lighten the color and control the shrinkage. Silica, alumina, ground biscuit, and calcined kaolin are used.

Stains. The stains are essentially the same as those used in overglaze colors — raw oxides or spinels.

Application. The ingredients, finely ground and washed of all soluble material,

are mixed with a vehicle such as an oil and turpentine, gum and water, or glycerine and water, for application to the biscuited piece. The underglaze color, in most cases, must be fired on at red heat to drive off the volatiles and oxidize the carbon in the vehicle. Otherwise bubbles would be created later in the glazing operation. In order to obtain an even glaze layer, it is desirable to have the porosity of the fired-on color about the same as that of the biscuit. After glazing and firing, the colors are strengthened by being imbedded in the glaze layer, in much the same way that a wet pebble taken from a brook is more brilliantly colored than it is when dry.

Stability of colors. Of course the firing temperature must be that of the glaze, so underglaze colors must be adapted to each type of ware. Earthenware may be fired at 1175°C (cone 6) while hard porcelain is fired up to 1500°C (cone 19). The range, and in general the brilliance, of the colors decreases with increasing temperature. At 1500°C, blues, grays, tans, and gold pinks are about the only colors available. In our own laboratory we have been trying to find a color stable at 1850°C (cone 40) to mark refractory ware, but as yet have found nothing suitable.

Lusters

This type of decoration was used at an early date in Persia and by the Moors. It consists of a thin, more or less iridescent layer of metal or oxide on the surface of a glaze. There are two basic types, one produced in a reducing fire and the other under oxidizing conditions. Lusters may be either colored or colorless.

The luster is prepared by forming a metal resinate and then mixing this with a vehicle, such as oil of lavender, for application to the glaze. The piece is then fired in a muffle at

600 to 900°C, whereby the resinate and vehicle are decomposed and the resulting carbon acts on the easily reducible oxide to produce the metal as a thin film on the glaze. It is probable that this operation takes place in the vapor phase, as the volatile bismuth is almost always present in a luster. Sometimes several lusters are applied one over the other to give special effects. For the reducing fire luster, the resinate is not used, but the metal salts, including bismuth, are applied with gum and water, and then fired in a muffle with a strongly reducing atmosphere.

Gilding

This is really an overglaze decoration consisting of a layer of a noble metal, usually gold, but sometimes silver or platinum.

Soluble gold process. In this process a soluble gold salt is incorporated in a varnish and applied to the glaze surface. On firing at about 700°C the gold is reduced to metal and deposited on the glaze in a thin layer in the manner of a luster. The gold, laid down in this way, is very thin and wears off quite easily so this method is used only on inexpensive ware.

Coin gold. In another method, powdered coin gold is mixed with a flux similar to those used for overglaze colors. Carried in a vehicle of oil, or water soluble gum, it is applied to the glazed piece. The firing is carried out at 700 to 800°C. After the firing the gold decoration looks brown with no metallic appearance. To form a polished gold surface the layer is burnished by rubbing with a smooth stone or, if a mat surface is required, with spun glass. This mechanical pressure causes the soft metal to flow into a continuous layer. This method of gilding is an expensive one and is only used on high priced ware. The layer is quite durable, but being soft can be scratched.

REFERENCES

ANDREWS, A. I., *Ceramic Tests and Calculations.* John Wiley and Sons, Inc., New York, 1928.

BINNS, C. F., *Manual of Practical Potting*, 5th ed. Scott, Greenwood and Son, London, 1922.

HAINBACH, R., *Pottery Decorating.* Scott, Greenwood and Son, London, 1924.

KOENIG, J. H., and EARHART, W. H., *Literature Abstracts of Ceramic Glazes.* Industrial Publications, Inc., Chicago, 1942.

KOHL, H., "Yellow Colors in the Ceramic Art." *Ceramic Age* **29**, 118, 1937.

MELLOR, J. W., "Cobalt and Nickel Colors." *Trans. Brit. Ceram. Soc.* **1**, 1936–1937.

NEWCOMB, R., JR., *Ceramic Whitewares.* Pitman Publishing Corp., New York, 1947.

PARMELEE, C. W., *Ceramic Glazes.* Industrial Publications, Inc., Chicago, 1948.

SEARLE, A. B., *An Encyclopedia of Ceramic Industries.* Benn, Ltd., London, 1929–1930, 3 vols.

CHAPTER 23

DECORATIVE PROCESSES

Introduction

There are many different processes used for decorating ceramic ware. There is space here to cover only the more important or more striking ones, but the student who is interested will find much literature on the subject. Visits to our potteries will allow him to see mass production methods of decoration in operation, while the European potteries may be counted on to illustrate the various hand decoration methods.

Modeling

Probably the earliest type of pottery ornamentation was relief carving on the surface; at first as a simple scratched design and later as sculptured reliefs.

Relief decoration. A freshly thrown vessel may have a relief modeled on the surface directly with some of the same clay. Early Roman and modern Italian pieces are typical. Usually, however, the modeling is in the surface of a mold, in which a succession of pieces may be formed by jiggering, molding, or casting. As an example, a plate with a raised border in relief may be taken. A model of the plate itself is turned from plas-

ter, and a shallow groove cut in the rim in the position of the relief band. This groove is then filled with wax, in which the relief is roughly modeled. A negative plaster cast is then made of the positive surface of the plate and further modeling done in the plaster. Alternate positive and negative casts are then made with modeling on each until the relief is perfected. The reason for doing this is very evident when one tries to model a delicate relief, for it is much easier to shape the humps than the hollows. The cross-sections in Fig. 23–1 will show the various steps in this process.

Applied reliefs. Early English sprig earthenware made in the middle of the 17th century produced relief decorations by forming wet clay in a small mold and at once pressing it, mold and all, onto the fresh surface of the vessel and then removing the mold. This rather crude method is not used at present, but the cameo ware, perhaps best known from the Wedgwood factory, is a refinement of this. Here the plastic body is pressed into small molds made of biscuit. After a short drying period the molded relief, often of a very delicate structure, is gently burnished

FIG. 23–1. Steps in modeling a relief.

203

on the back with a spoon which stretches the body slightly and allows it to be released from the mold; then it is temporarily set on a wet plaster block to prevent its drying. Now the surface of the piece is moistened with water and the relief section carefully applied. As the water dries, it draws the relief down into good contact. Obviously great skill is needed in this process.

An entirely different method of forming reliefs is the *pâte-sur-pâte* (paste on paste) method brought to a high degree of perfection in both England and France. As each piece is the individual work of the artist, the better pieces are in great demand by collectors. The process consists of building up a relief, usually in white porcelain slip on a colored background of raw, wet body. This is done with a brush, and layer after layer is added until a very low relief is completed. On firing, the varying translucency gives a most delicate effect, so that draperies may be superbly rendered. Often a clear glaze covers the whole. Although much inferior ware may be seen, a few artists have turned out exquisite pieces.

Sculpture in the round. Much sculpture has been produced in ceramic materials, from huge terra-cotta pieces to tiny figurines. In general, two methods are employed: the first is direct modeling in the body; the second consists of making a model, then a mold from it, and forming the body in the mold by pressing or casting. The direct method is generally confined to rather simple pieces where only one of a kind is needed. An example would be a portrait head. In this case the sections would be rather thick and it would be necessary to hollow out the center in order to dry and fire safely. Another example of direct modeling is a figurine, such as that shown in Fig. 23–2.

In most cases, however, production of a number of pieces of the same kind is necessary, so a model is made in wax, plasticine, or lead. If this is complicated, it is cut up into an appropriate number of pieces and a plaster piece mold made from each one. Each piece mold is then filled by casting or pressing and the units thus produced are assembled with slip into the pose of the original model. Some of the figure groups may have as many as 30 pieces that must be assembled with great precision. In the case of fine ware the parting lines are suppressed as described in Chapter 11. The biscuit porcelain of Sèvres and the Parian porcelain of England, made in the form of figurines, are excellent examples of this process.

Printing methods

Early decorated pottery was hand painted and thus was too costly for the low priced market. Therefore, the invention of Sadler and Green of Liverpool, England, in the year 1756, whereby they were able to apply a printed decoration, was most important. This process, and it is still used today for some types of ware, consisted of coating an engraved copper plate with a special ink containing a ceramic stain, and transferring this ink to a thin paper which could then be placed on a piece of ware and rubbed down to transfer the ink again onto the body or glaze. The invention consisted fundamentally of transferring a design from a flat printing surface to a curved pottery surface by the intermediate step of using a flexible membrane. This principle is inherent in most modern mass production processes of ceramic decoration.

Decalcomanias. This type of decoration has been very extensively used in the past because it is inexpensive and permits multiple colors. While this method has been displaced to some extent by the silk screen

FIG. 23–2. The Town Crier. Direct modeling by a pupil of Forsyth, Stoke-on-Trent, England.

process, it is still important where several colors are desired.

The decalcomanias, or *decals*, are printed on special duplex paper, made with a heavy backing lightly attached to a thin tissue face that is coated on the outside with a soluble sizing, by the normal lithographic or offset printing process using up to eight colors. The inks are made up with overglaze stains and a waterproof vehicle, such as a varnish, that hardens on the paper. In the dust process, varnish only is printed on the paper, and the dry ceramic stain dusted onto it. The printing is done on large sheets, usually 45 inches by 29½ inches, containing all the patterns for one of each shape in a set, which are then cut up to give a number of units for smooth application to the ware.

The decals are applied by first coating the glazed ware with varnish, letting this become tacky, and then rubbing the decal tissue with the backing paper removed, face down into this varnish. The ware is immersed in water, which floats off the decal tissue and leaves the ink in the varnish layer. The ware can then be fired like any overglaze decoration.

Silk screen process. This process has come into extensive use during the 1940's, to some extent displacing decals. It has proved particularly useful for decorating glassware.

The process essentially consists of a silk bolting cloth of 125 to 150 meshes per inch coated with a stencil film, except in those areas where a pattern is desired. In other words it is a reinforced stencil. The stencil may be prepared by cutting out areas of a thin decalcomania tissue mounted on a heavy backing. This tissue is then mounted on the stretched silk screen with shellac and a hot iron, after which operation the backing is stripped off. If the design has fine detail, the photographic process is used, in which the stretched silk screen is coated with gela-

tine that has been sensitized with potassium bichromate. This is exposed through a negative to a strong light, and washed in water. The unexposed portions will be soluble and leave clear areas on the silk for printing.

In the actual printing operation the silk screen is placed a short distance above the surface of the ware. A rubber squeegee blade then moves across the silk screen, forcing it down against the ware, and at the same time pressing a viscous ink down through the holes in the open areas to form the pattern. Automatic machines have been developed to print by this method, but there is not space here to go into their design. If more than one color is required, each ink must be dried and a second printing made, registering with the first one.

Photographic method

This method has many attractive possibilities, but as yet has not been used very extensively. Most of the processes consist of coating the ware with a light-sensitive, dichromate gelatine and exposing it through a negative. The unexposed portions are washed off, leaving a graded film which may contain coloring pigments to be fired on later.

Other processes

Ground laying. This interesting process is used on fine china to provide an even color over large areas. The glazed piece is carefully painted with a solution of sugar and a dye over all areas not to be covered by the ground. After drying, an oil (linseed oil and turpentine) is brushed on and then patted down with a soft pad to eliminate brush marks. Next, the dry, powdered overglaze color is dusted into the oil with a piece of cotton and the excess blown off. After drying about 24 hours, the piece is washed in water which removes the areas painted with sugar solution. The pieces are then fired in the usual way for overglaze color and produce a ground that is uniform both in color and texture.

Hand decoration. Much fine ware is decorated by hand with brushes or needle sprays. This technique requires much experience and the European potteries have set up training schools for their apprentices. In this country little hand decoration is used, because of high labor costs, lack of trained personnel, and the reluctance of the American public to buy domestic production in the higher priced classification.

Hand decoration may be applied under the glaze or over the glaze, as it is for the printing processes. The same ceramic colors are used, but the medium may differ. Some colors are applied with water soluble gums or glycerine, while others are applied with oils and a thinner like turpentine. A variety of brushes and application techniques are used which cannot be detailed here.

Considerable hand decoration is executed with the air brush, particularly for underglaze application. It is possible to spray on an area of color and then erase small spots to give a white design. For example, a

Fig. 23-3. Porcelain fawn with underglaze decoration. Made at Copenhagen by the Royal Danish Pottery.

Royal Danish Pottery fawn, Fig. 23–3, is made by spraying on a brown underglaze, then with a pointed rubber erasing the color to form the white spots. The advantage of the spraying method is the very uniform surface produced.

Stencil decoration. Simple designs may be readily achieved by spraying through stencils. These stencils are made of thin metal and are held a short distance from the ceramic surface. By proper registration of the stencils several colors may be applied to the same piece.

Stamping. Color may be applied with rubber stamps cut to the desired outline. Because of the distortion of the rubber, it is difficult to obtain clean, accurate application.

Sgraffito decoration. This is one of the simplest of methods. The green piece is coated with a thin layer of slip of a color different from the body. After the piece has dried, the outer coat is scraped away, to show the body in selected areas. The piece is then fired, and sometimes glazed.

Inlaying. Some of the early floor tiles in the English cathedrals have inlaid decorations that have withstood centuries of wear. This process is carried out by pressing colored clay into a mold so that the design is recessed below the surface. After the tile has dried slightly, another colored clay is pressed into this recess and scraped off flush. The whole is then burned to make a durable structure.

Encaustic decoration. This is a glaze decoration in enamel colors on tile. To separate the various colors so that they will not diffuse into each other, three methods are used. In the first, ridges are molded in the tile to define each area; in the second, grooves are used; and in the third, a narrow band of underglaze black. A heavy layer of glaze is flowed into each section by means of a syringe, and the whole fired. The effect is much like stained glass on a small scale.

Banding. Bands are put on plates by revolving them on a wheel and holding a brush against them at the proper place. This manual operation required great skill and long training. Now, banding machines have been designed to put on bands automatically at a great saving in labor costs.

REFERENCES

"Applying Bands and Lines on Greenware." *Ceramic Industry*, October, 1949, p. 87.

Binns, C. F., *Manual of Practical Potting*, 5th ed. Scott, Greenwood and Sons, London, 1922.

"Copper Plate Printing." *Ceramic Industry*, February, 1950, p. 75.

Cox, W. E., *The Book of Pottery and Porcelain*. Crown Publishers, New York, 1944.

"Lining Overglaze." *Ceramic Industry*, December, 1949, p. 83.

"Lining Underglaze." *Ceramic Industry*, November, 1949, p. 86.

"Making Printing Oil." *Ceramic Industry*, March, 1950, p. 87.

Newcomb, R., Jr., *Ceramic Whitewares*. Pitman Publishing Corp., New York, 1947.

Searle, A. B., *An Encyclopedia of Ceramic Industries*. Benn, Ltd., London, 1929–1930, 3 vols.

CHAPTER 24

CEMENTS AND PLASTERS

Introduction

Cements are of a great variety of compositions but have in common the property of setting without heat. They are used for many purposes, varying from the production of a massive structure damming a river, to the filling of a tooth cavity. Here it will be possible to consider only the more important types.

Lime

Calcination. The raw materials for lime are limestone (Calcite, $CaCO_3$) or oyster shells (Aragonite, $CaCO_3$) which on calcination become quick lime, or CaO. This reaction starts at about 900°C and is carried out in shaft or rotary kilns. The oxide is very reactive and may be slaked with water to form hydrated lime [$Ca(OH)_2$] with the evolution of considerable heat.

Mortars. The lime hydrate or lime putty is mixed with sand to form a mortar. This mixture slowly sets by absorbing CO_2 from the air to form $CaCO_3$, as well as by reacting with the silica to form calcium silicate. Lime mortars have been much used in the past for laying brick. In spite of their slow set and low mechanical strength, they formed excellent watertight walls, since their plasticity produced well-filled joints. At present, mortar is made by adding Portland cement to the mixture for more rapid set and greater strength. A typical mortar for brick setting above grade would be:

1 volume Portland cement
2 volumes lime hydrate
8 volumes sand

Portland cement

Portland cement is one of the hydraulic cements that sets in the presence of water. It is one of our most important structural materials, having good compressive strength and reasonable durability. However, it is a very basic material and, from the point of view of the geologist, is not an ideal composition to withstand weathering.

Manufacture. Portland cement has a composition as follows:

Silica	19–25 per cent
Alumina	5–9
Ferric oxide	2–4
Lime	60–64
Magnesia	1–4
Sulphur trioxide	1–2

In Fig. 24–1 is shown a three-component diagram of lime-silica-alumina with the area

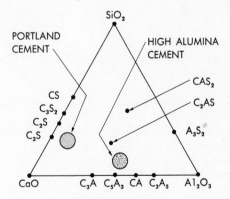

Fig. 24–1. Composition of Portland cement in the $CaO-Al_2O_2-SiO_2$ system in weight per cent. The composition of high alumina cement is also shown.

of Portland cement composition indicated when ferric oxide is combined with the alumina, and magnesia with the lime. On the same diagram is also shown the area for high alumina cements, which will be discussed later.

The composition for Portland cement may be obtained by combining clays and limestone in the correct proportion, or it may be made of corrected slags from the steel industry. There are two general processes used, the wet and the dry. In the former the raw materials are wet ground in ball mills to a slurry for feeding the rotary kiln. In the latter the grinding is dry and the powder is fed into the kiln.

The burning is carried out in large rotary kilns with inside diameters up to 10 feet and lengths up to 250 feet. As the raw material slowly moves down the revolving kiln, it en-

counters hot gases moving in the opposite direction and so is heated up to around 1600°C when it reaches the hot end. It is then discharged in the form of clinker nodules. The chief reactions occurring in firing are shown in Fig. 24–2. The firing is usually accomplished by means of a powdered coal burner, and the waste heat is recovered in a boiler at the other end. The kiln is lined with refractory material, dense fire clay at the cool end and either a high alumina brick or a magnesite brick at the hot end. Good operation depends on building a layer of clinker on the brick surface. A cross section of a kiln is shown in Fig. 24–3. The clinker is cooled in a rotary cooler below the kiln with a recovery of the heated air for the burner. The clinker is then ground in ball mills and bagged. A flow sheet for the dry process is shown in Fig. 24–4.

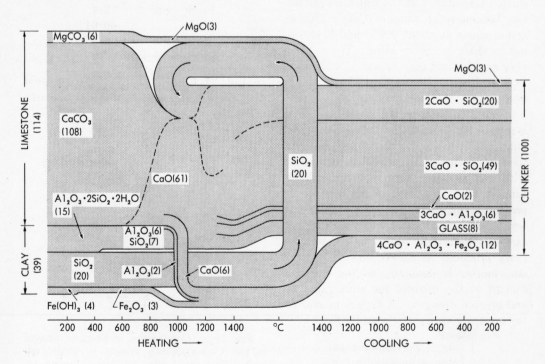

FIG. 24–2. Reaction occurring during Portland cement clinker formation.

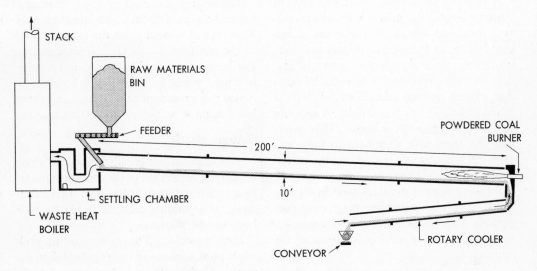

FIG. 24–3. Cross section of a rotary kiln for burning Portland cement clinker.

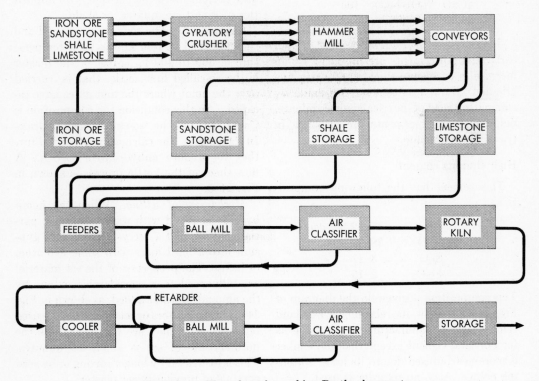

FIG. 24–4. Flow sheet for making Portland cement.

Mechanism of setting. Portland cement contains largely the minerals tricalcium silicate, dicalcium silicate, tetracalcium aluminum ferrite, and tricalcium aluminate. On hydration, there is solution, recrystallization, and the precipitation of colloidal silica that causes setting. As Portland cement would ordinarily set too rapidly, a retarder of 3 per cent gypsum is used. This reacts with the tricalcium aluminate to form calcium sulphoaluminate that slows the set and decreases the shrinkage.

Portland cement develops heat in setting, which in massive structures may cause dangerously high temperatures unless cooling is used. The heat of setting for some of the minerals is:

$3CaO \cdot Al_2O_3$	207 cal/gm
$3CaO \cdot SiO_2$	120
$4CaO \cdot Al_2O_3 \cdot Fe_2O_3$	100
$2CaO \cdot SiO_2$	62

Special types of Portland cement. High early strength cements are produced by an increase in tricalcium silicate and extra fine grinding. Low heat cements are made by eliminating most of the tricalcium aluminate. Sulphur resisting cements are also low in tricalcium aluminate.

High alumina cement

This cement has the following composition:

CaO	35–42 per cent
Al_2O_3	38–48
Si_2O	3–11
Fe_2O_3	2–15

This composition is given on the diagram of Fig. 24–1 to show its relation to Portland cement. The manufacture is the same as for Portland cement except that the clinker is more nearly fused due to its lower softening point. Also, no retarder is used.

The setting is caused by the formation of hydrated alumina from tricalcium aluminate. This type of cement gains the same strength in twenty-four hours that is attained by Portland cement in thirty days, and is therefore used where quick setting is helpful. It is also more resistant to sea water and shows less fluxing in refractory air setting mortars.

Gypsum plaster

This material, often called plaster of Paris, is not only important to the ceramic industry as a mold material, but is also used extensively as a building material in blocks, wallboards, and plasters.

Raw material. The raw gypsum rock, when pure, consists of hydrated calcium sulphate, $CaSO_4 \cdot 2H_2O$. There are deposits containing 99 per cent gypsum, but others may drop as low as 65 per cent with impurities of limestone, quartz, or shale.

Calcination. The rock is fine ground and then calcined in iron kettles at a temperature of 300 to 400°F. Most of the plaster is the so-called first settle, that is, carried over the point where the first gases have escaped. In this condition the composition is $CaSO_4 \cdot \frac{1}{2} H_2O$, the so-called hemihydride. In a few cases, the calcination is carried further to produce anhydride or $CaSO_4$. A flow sheet of the kettle process is shown in Fig. 24–5.

Mechanism of setting. When the hemihydride is mixed with water there is a partial solution and a recrystallization of gypsum into a hard mass with some evolution of heat. The properties of the set material are influenced to a considerable extent by the amount of water added, as shown in Fig. 24–6. Accelerators or retarders may be used but ordinarily they are not needed, as the normal time of set is about 20 minutes. This is decreased by longer mixing or by seed crystals of previously set plaster.

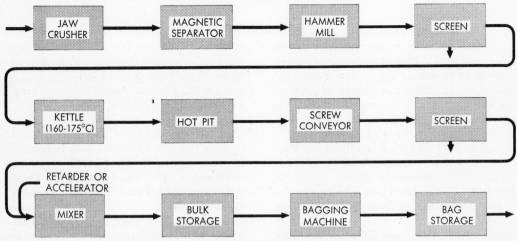

FIG. 24–5. Flow sheet for making gypsum plaster.

Special types of plaster. Hard gypsum plasters are made with greater strength and hardness than ordinary potters' plaster. These are useful for models and pressing molds.

Oxychloride cements

Magnesium oxychloride cements, often called Sorel cements, are made by mixing caustic MgO with a $MgCl_2$ solution. Magnesium oxychloride is slowly formed to produce a hard structure. These cements are used for interior floors. Zinc oxychloride cement was used for dental purposes but at present has been displaced by other types.

Silicate cements

These cements are made from silicate of soda and a filler such as fine ground quartz. On drying, the mass becomes hard, and on heating slowly it may be stabilized. These cements are used for acid resisting construction, since acid precipitates silica gel, which is then inert.

Silicate of soda with alumina and a fluoride have been used for dental purposes. Silica and alumina gels are formed with the valuable property of translucency.

Phosphate cements

These cements have been used for dental purposes. The oxides of Zn, Zr, or Cu are mixed with phosphoric acid, which reacts to form the metaphosphates.

For air setting refractory purposes the phosphate cements look promising because of their strength at all temperatures.

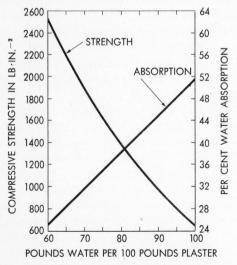

FIG. 24–6. Properties of gypsum plaster with varying amounts of mixing water.

REFERENCES

BOGUE, R. H., *The Chemistry of Portland Cement.* Reinhold Publishing Corp., New York, 1947.

BOGUE, R. H., "The Constitution of Portland Cement Clinker." *Concrete,* July, 1926, to February, 1927.

BOGUE, R. H., "A Digest of the Literature on the Nature of the Setting and Hardening Process in Portland Cement." *Rock Products,* May, 1928, to September, 1928.

ECKEL, E. C., *Cements, Limes, and Plasters,* 3rd ed. John Wiley and Sons, Inc., New York, 1928.

KELLEY, K. K., SOUTHARD, J. C., and ANDERSON, C. T., *Thermodynamic Properties of Gypsum and Its Dehydration Products.* U. S. Bur. of Mines, Tech. Paper No. 625, 1947.

MEADE, R. K., *Portland Cement,* 3rd ed. Chemical Publishing Company, Easton, Penna., 1926.

OFFUTT, J. S., and LAMBE, C. M., "Plasters and Gypsum Cements for the Ceramic Industry." *Bull. Am. Ceram. Soc.* **26,** 29, 1947.

ROBITSCHEK, J. M., "The Cement's the Thing in Ceramic Acid Tanks." *Ceramic Industry* **42,** 31–35, 1944.

ROBITSCHEK, J. M., "Right Cement Makes for Better Acid Proof Structures." *Ceramic Industry* **42,** 38, 1944; **42,** 88, 1944.

APPENDIX

Kilns

A number of small kilns for use in the laboratory or by the studio potter are described in this section. They are all of simple construction and have been thoroughly tried out. The kiln is always the stumbling block for the beginner in ceramics, as attested by the hundreds of letters the author receives every year asking for information on this subject. There are many excellent kilns on the market, but naturally they are more expensive than those made at home. In addition to the cost factor, there is a certain satisfaction to be had from making one's own equipment.

In selecting a kiln, a decision must be made as to the maximum temperature needed and the size of the firing chamber.

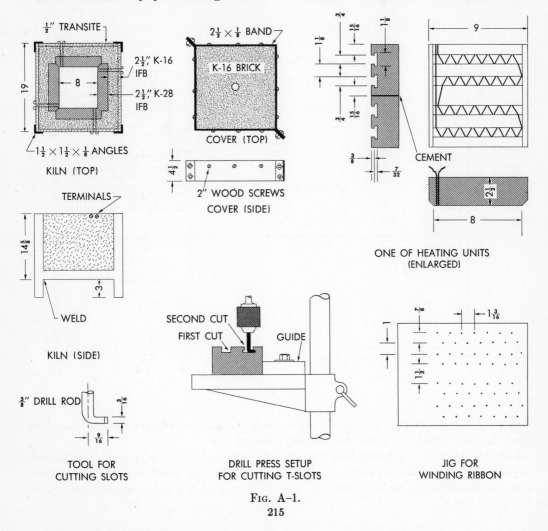

Fig. A–1.

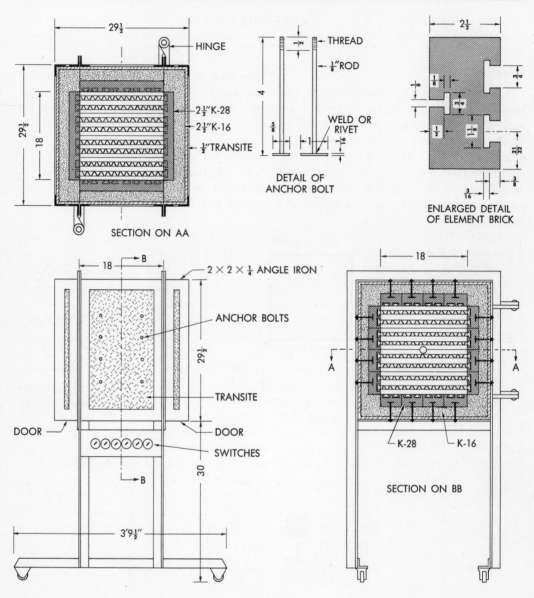

Fig. A–2.

The cost of the kiln goes up rapidly with both factors. Next, it must be decided whether an electric or gas-fired kiln is most applicable to the particular conditions. Before an installation is made, of course, the local fire and zoning regulations should be looked into.

Small electric kiln for medium temperatures. This little kiln, which has a working space of 8 by 8 by 9 inches, may be used on an ordinary 110-volt house circuit. It requires only 1200 watts, the same as a toaster. Nevertheless, it provides a uniform temperature as high as 1165°C (cone 4) and may be used for test work or pottery firing.

The construction should be clear from Fig. A–1. The four heating elements form the sides of the kiln and are easily removed for the purpose of replacing burned out windings. Each element is made of two straight insulating firebrick cemented together with air-set mortar, edge to edge, and grooved with T-slots as shown. The insulating firebrick is an ideal material for small kilns, since it may be cut and formed with great ease, is light in weight, and is an excellent insulator. The slots are cut by first making a groove $\frac{3}{8}$ in. wide and $\frac{9}{16}$ in. deep with an old router blade, and then undercutting with a tool shown in Fig. A–1. This work can be conveniently done on a drill press, using a guide block. The 85 inches of 0.01 by $\frac{3}{16}$ in. Kanthal A–1 ribbon is wound on a jig as shown, and then slipped into the slots.

The cover is made by cementing together insulating firebrick and clamping them in a metal band. A sight hole may be left in the center and covered with a small piece of brick.

The materials needed for this kiln are:

2 pieces $\frac{1}{2}$-in. transite,[1] $18\frac{7}{8}$ by $13\frac{1}{2}$ in.
2 pieces $\frac{1}{2}$-in. transite, $17\frac{7}{8}$ by $13\frac{1}{2}$ in.
1 piece $\frac{1}{2}$-in. transite, $18\frac{7}{8}$ by $18\frac{7}{8}$ in.
12 ft of 2 by 2 by $\frac{1}{8}$ in. angle iron
6 ft of $4\frac{1}{2}$ by $\frac{3}{32}$ in. strap iron
30 ft of 0.01 by $\frac{3}{16}$ in. Kanthal ribbon[2]

1 carton (25) K–28 insulating firebrick[3]
1 carton (25) K–16 insulating firebrick[3]
10 pounds air setting mortar[3]

The cost of these materials in 1951 was around $40.

Wire-wound electric box kiln. This kiln, shown in Fig. A–2, has the same general construction as the preceding one, but it has eight times as large a cubic capacity. This is an excellent kiln for the serious potter. We have used one in our laboratory for fifteen years. Many have been built by research laboratories, as well as by advanced craftsmen. Setting is simplified by the two doors, and the temperature is remarkably uniform because of the presence of heat on all sides. Although this kiln has been used up to 1260°C (cone 10), the life of the winding under these conditions is short; a maximum temperature of 1165°C (cone 4) is recommended for long-continued use.

The construction should be clear from the drawing. The important features of this kiln are the tie bolts holding the lining brick, the continuous heating surface, and the two full-size setting doors. The heating unit bricks are cut in the same way as for the previous kiln, except that there is an additional T-slot in the back. The winding, 400 inches of 0.3 by 3 mm Kanthal A–1 ribbon for a whole side, is made up on a jig and slipped into the grooves. The ends of the winding are brought out to brass connecting screws mounted on the transite. From here flexible cable connects them to the switches.

The anchor bolts are made as shown in Fig. A–2 from 18–8 stainless steel rod and

---

[1] H. W. Johns-Manville Co., New York, N. Y.
[2] C. O. Jelliff Mfg. Corp., Southport, Connecticut.

[3] Babcock and Wilcox Co., 85 Liberty Street, New York, N. Y.

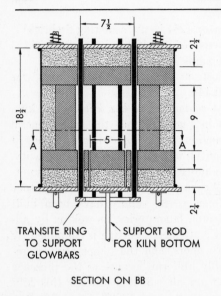

TRANSITE RING TO SUPPORT GLOWBARS   SUPPORT ROD FOR KILN BOTTOM

SECTION ON BB

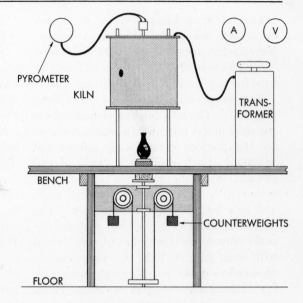

PYROMETER

KILN

TRANS-FORMER

BENCH

COUNTERWEIGHTS

FLOOR

A CONVENIENT MOUNTING FOR THE KILN WITH BOTTOM LOWERED

K-16 IFB

K-28 IFB

SIGHT HOLE

SECTION ON AA

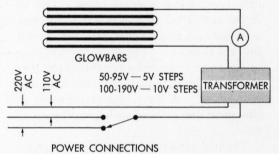

GLOWBARS

220V AC   110V AC   50-95V — 5V STEPS
100-190V — 10V STEPS   TRANSFORMER

POWER CONNECTIONS

FIG. A-3.

sheet, spot-welded or riveted together. For low temperature use ordinary steel is suitable for this purpose. Each section is connected to a separate switch, so that from one to six sections may be used at a time to give some temperature regulation. They are all in parallel on 110 volts or two in series and three in parallel on 220 volts, using a maximum of 8 KVA. While switching sections on and off gives a fair degree of control, it is preferable to supply the kiln with a 10 KVA autotransformer with one 50-volt step, three 20-volt steps and three 5-volt steps.

The materials needed for this kiln are:

100 ft of 2 by 2 by $\frac{1}{4}$ in. angle iron
4 special hinges
6 — 10 amp, 2-pole snap switches
60 stainless steel anchor bolts
4 heavy casters
$\frac{1}{2}$-in. transite sheet as follows:
2 pieces, $27\frac{7}{8}$ by $27\frac{7}{8}$ in.

2 pieces, $27\frac{7}{8}$ by 18 in.
2 pieces, $28\frac{7}{8}$ by 18 in.
4 pieces, 5 by $27\frac{7}{8}$ in.
4 pieces, 5 by $28\frac{7}{8}$ in.
210 ft of Kanthal A–1, 3 by 0.3 mm ribbon
12 — $\frac{1}{4}$ by 1 in. brass bolts
100 (4 cartons) K–28 insulating firebrick
100 (4 cartons) K–16 insulating firebrick
4 — $17\frac{1}{2}$ by $17\frac{1}{2}$ by $\frac{1}{2}$ in. silicon carbide shelves[1]
48 special anchor bolts

The total cost of these materials in 1951 was about $250.

Small electric globar kiln. This is a small kiln, shown in Fig. A–3, for experimental work up to 1500°C (cone 20). It has the great advantage of permitting examination of the ware at any point in the firing cycle by momentarily dropping the bottom. It is particularly suited for work with porcelains and crystalline glazes.

This kiln must be supplied by a variable voltage from a 5-KVA autotransformer, with the steps shown in Fig. A–3, or by a continuously variable transformer, such as a Variac. The maximum voltage to the transformer is 220 and the current is 20 amperes. The globars are connected initially in series. The lower ends of the globars are supported by a transite ring spaced below the kiln bottom with three bushings. The globars should have a $\frac{3}{4}$-in. clearance hole through the insulating brick and a $\frac{5}{8}$-in. hole through the transite. It is important that there be no binding. The ends of the globars are connected to brass screws by means of the regular terminal straps and clips supplied with them.

The material required for this kiln is as follows:

1 sheet iron shell
2 — $\frac{3}{4}$-in. transite sheets, 21 by 21 in.
1 — $\frac{1}{2}$-in. transite disk, 6-in. dia.
1 — $\frac{3}{8}$-in. transite disk, $8\frac{1}{2}$-in. dia.
25 (1 carton) K–28 insulating firebrick
25 (1 carton) K–16 insulating firebrick
6 globars 21 by 10 by $\frac{1}{2}$ in.[2]
12 terminal clips[2]
12 flexible terminal straps[2]
12 brass bolts, $\frac{1}{4}$ by 1 in.
$\frac{1}{2}$-in. pipe, pulleys and counterweight
5-KVA variable voltage transformer

These will cost at 1951 prices about $75, exclusive of the transformer.

Large electric globar kiln. This kiln, as shown in Fig. A–4, is more expensive than the preceding one, but has a much greater capacity. At lower temperatures (1300°) the life of the heating units is quite long, up to 100 firings. This kiln requires a special autotransformer of 20-KVA capacity and having voltage steps like those shown in Fig. A–4.

The materials for the kiln consist of:

12 flexible terminal straps
12 terminal clips
6 — 39 by 22 by $\frac{3}{4}$ in. globars
65 ft of $2\frac{1}{2}$ by $2\frac{1}{2}$ by $\frac{1}{4}$ in. angle iron
$\frac{1}{2}$-in. transite:
　1 piece $37\frac{7}{8}$ by $38\frac{7}{8}$ in.
　1 piece $39\frac{1}{2}$ by $38\frac{1}{2}$ in.
　4 pieces $35\frac{1}{2}$ by $37\frac{7}{8}$ in.
4 casters
2 — 25 by 23 by $\frac{1}{2}$ in. silicon carbide plates
4 — $22\frac{1}{2}$ by 16 by $\frac{1}{2}$ in. silicon carbide plates
250 (10 cartons) K–28 straight brick
150 (6 cartons) K–16 straight brick

At 1951 prices this material will cost about

---

[1] A number of manufacturers of kiln furniture advertise in *Ceramic Age* and *Ceramic Industry.*

[2] Globar Corporation, Niagara Falls, N. Y.

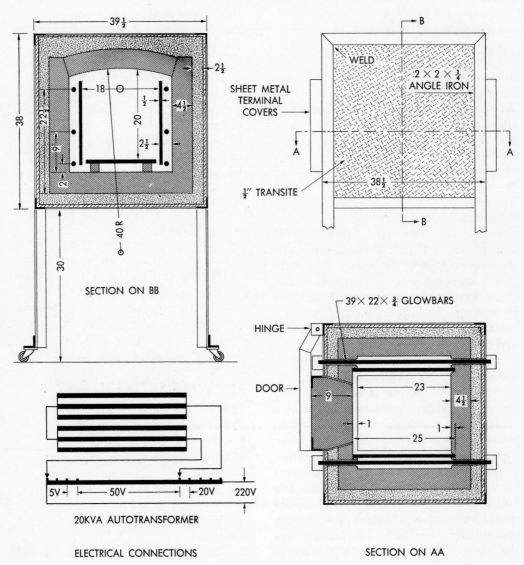

FIG. A–4.

$300. The transformer will cost considerably more than this when the wiring is included.

Gas-fired muffle kiln. A kiln of this type has been used almost daily in our laboratory for nearly twenty years. The only repairs needed have been two replacements of the muffle liner. Temperatures up to 1400°C (cone 15) can be reached, a heat which permits firing porcelain. The burners are laboratory Méker type, allowing good control of both temperature and atmosphere. The design of the gas passages is such that a uniform temperature can be attained from

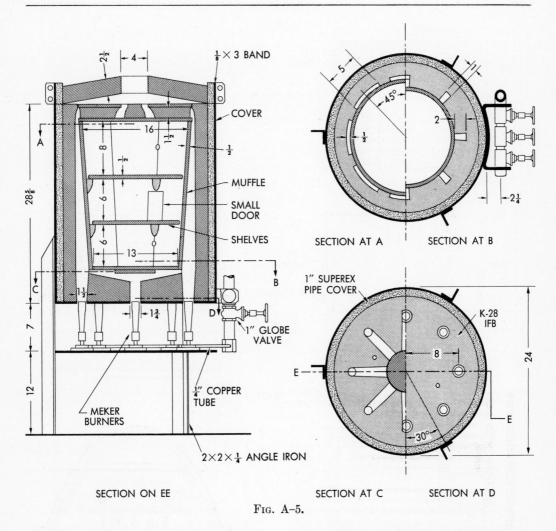

SECTION ON EE

FIG. A–5.

top to bottom; however, care should be taken to follow the drawing exactly and center the burners in the tuyères. A great many kilns have been built according to the drawings of Fig. A–5 with excellent success. No direct chimney connection is necessary, but in a low-studded room a collecting hood is desirable. This kiln may be constructed with five burners and four muffle flues if temperatures of no more than 1225°C (cone 8) are needed.

The following material is required to construct the kiln, which will cost about $150 (1951).

1 cover band
1 steel shell
10 ft of 2-in. angle iron
15 pieces of 1 by 6 by 36 in. Superex Block[1]
5 round kiln plates (SiC) $\frac{1}{2}$ in. thick (one each of 17, 14, 13, $12\frac{1}{2}$, and 6-in. dia.)

---

[1] H. W. Johns-Manville Co.

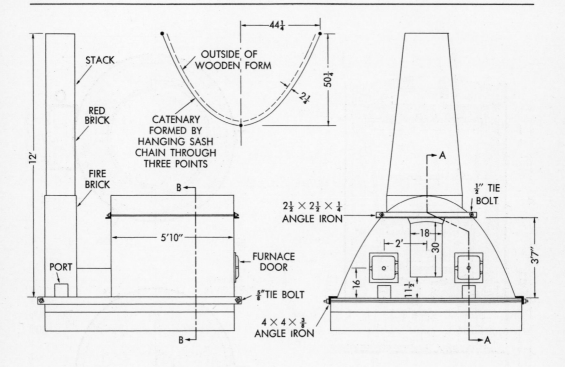

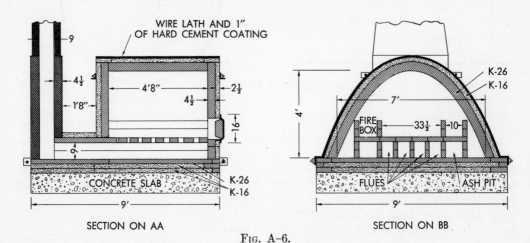

Fig. A–6.

24 muffle staves[1]
100 B & W K–28 brick (4 cartons)
9 Méker burners

3 gas valves
18 ft of $\frac{1}{4}$-in. copper tubing
1-in. pipe fittings

---

[1] These may be formed from refractory body No. 10 of the next section.

Wood-fired pottery kiln. This kiln shown in Fig. A–6 was designed and built by

Table A-1

Summary of Kilns

Kiln	Capacity in cu ft	Cost of firing[1] to			Cost of materials (1951)
		1100°C	1300°C	1500°C	
Small electric wire-wound	0.3	$0.35	--	--	$40.00
Large electric wire-wound	3.4	$2.50	--	--	$250.00
Small electric globar kiln	0.1	$0.40	$0.70	$1.50	$75.00
Large electric globar kiln	3.4	$6.00	$12.00	$20.00	$300.00
Gas-fired muffle kiln	1.8	$0.50	$0.70	$1.40	$150.00
Wood-fired kiln	36	$12.00	$24.00	--	$1000.00

[1]Electricity figured at 5 cents per kwh, gas at $1.00 per 1000 cu ft (520 Btu), and wood at $24.00 per cord.

William M. Shakespeare to give efficient firing. If small dry wood is used, a temperature of 1125°C (cone 1) is easily reached. The ware should be placed in saggers, and the kiln fired for about twelve hours. This is a much shorter time than usual, because of the lightweight lining. About a quarter of a cord of wood is required, and of course, constant attention is needed over the firing period. The same kiln can be fired with gas or oil, although no performance data are available for these fuels.

The following materials are required, costing in 1951 about $1,000.

Common red brick. 3000
Low heat duty straight firebrick. 1200
Low heat duty soap firebrick. 50
Low heat duty No. 3 arch firebrick. . 15
Low heat duty No. 2 arch firebrick. . 15
Low heat duty $2\frac{1}{2}$ by $4\frac{1}{2}$ by $13\frac{1}{2}$ fire-
brick. 50
Insulating firebrick (B & W, K-28). . 1000
Insulating firebrick (B & W, K-16). . 500
Insulating block 1 in. thick, sq ft. . . . 150
Sheets of steel lath. 11
Mortar for firebrick, lb. 600

Mortar for ins. firebrick, lb. 500
Mortar for ins. block, lb. 200
Mortar for red bricks, lb. 500
Portland cement, bags. 17
Sand, cu yd. . . . . $3\frac{1}{2}$
Crushed stone, cu yd. . . . . $3\frac{1}{2}$
Broken stone, tons. 3
4 by 4 by $\frac{3}{8}$ angle iron, ft. 20
$2\frac{1}{2}$ by $2\frac{1}{2}$ by $\frac{1}{4}$ angle iron, ft. 10
$\frac{5}{8}$-in. iron rod, ft. 20
$\frac{1}{2}$-in. iron rod, ft. 12
16 by 16-in. (frame) lined furnace doors   2

It will be noticed that the arch of the kiln is in the form of a catenary to give stability. This shape may be determined most readily by hanging a light chain on the face of a vertical wall through the required three points, as shown in the upper part of Fig. A-6. The wooden form for laying the arch is then made $2\frac{1}{4}$ inches inside this curve.

Summary. Table A-1 tabulates the economic characteristics of the six kilns described in this section. Both the cost of the kiln and the cost of firing increase rapidly with size and firing temperature.

Table A-2

Bodies and Glazes

Group	Temperature level	Body			Glaze		
		Body type	Fired color	Forming methods	Type	Texture	Used with body no.
I	1000-1020°C (Cone 06-05)	(1) Earthenware	Dark red	Throwing, modeling	Raw lead Raw lead Raw lead	Bright Opaque Mat	1, 2 2, 2, 5 1, 2, 4
II	1145-1165°C (Cone 3-4)	(2) Porcelain	Light cream	Casting	Fritted	Bright	6, 7
		(3) Parian (4) Terra cotta (5) Stoneware	White Dark pink[1] Red[1]	Casting Pressing Throwing, modeling			
III	1225-1250°C (Cone 8-9)	(6) Stoneware	Gray	Throwing, modeling	Porcelain	Bright	8
		(7) Semivit. whiteware	White	Throwing, modeling, casting	Porcelain	Mat	8
		(8) Porcelain	White	Casting			
IV	1450°C (Cone 16)	(9) Porcelain	White	Casting	Porcelain	Bright	9
		(10) Refractory	Cream	Pressing	Porcelain	Mat	9
	Special				Fritted	Crystalline	7

[1]Buff-colored with some brick clays.

Bodies and glazes

The beginner in the field of pottery-making, or even the more advanced student, is usually at a loss to know just which bodies and glazes are best suited to his purpose. There are thousands of formulas published, but the experimenter usually comes up against three major problems: first, these compositions mature at many temperature levels; second, important details of formulation are often omitted, such as grinding time, thickness of application, or means of deflocculation; and third, there is often no indication that a certain glaze can be used with a certain body without crazing.

An attempt is made here to describe a series of bodies, either vitreous or sufficiently dense to be leak-proof, and a series of glazes to go with them, all adjusted to four firing levels.

These bodies and glazes are listed in Table A-2. In selecting bodies and glazes, the available firing equipment must be an important factor. Group I can be fired in the least expensive wire-wound kilns to give sound ware, but neither white nor vitreous bodies are possible. Group II can be fired in the more expensive wire-wound kilns, particularly those described in the preceding section, or in gas-fired kilns. Groups III and IV may be fired in electric globar-heated kilns or in high firing gas kilns, such as the muffle kiln previously shown.

In describing the individual bodies and glazes, sufficient details are given so that even the beginner in this field should be able to turn out good ware and thus avoid the many trials and disappointments usually accompanying such an undertaking.

Bodies. Table A-3 describes the ten bodies of Table A-1. The treatment is normal in all cases. It is convenient to filter small batches in a large funnel, using filter paper on a perforated plate, and accelerating the process by vacuum from a sink pump. Body 1 filters very slowly and is better dried out in a pan. Wedging may be done by hand, but a small pan or auger is helpful. If a magnetic separator is available, the

Table A-3

Body Composition

Body number	1	2	3	4	5	6	7	8	9	10
Body type	Red earthenware	Low fired porcelain[1]	Parian	Terra cotta	Red stoneware	Gray stoneware	Semivitreous whiteware	Porcelain[2]	High fired porcelain[2]	Refractory body
Maturing temperature	1000-1020°C	1145-1165°C					1225-1250°C		1450°C	
Cone	02-03	3-4					8-9		16	
Potter's flint	90	60			290	225	340	190	220	
Potash feldspar		540	600				140	370	300	
Nepheline syenite					600					
Hanover stoneware clay						475				
Glacial brick clay	800			100	100					200
Klondyke crude kaolin		80					120			
Edgar plastic kaolin		80	150	80			120	70	80	
Edgar N.C. kaolin		80						220	280	
Lundy N.C. kaolin		80	150							
Bell's dark ball clay	100	80	80				100			
Ky. Old Mine No. 4 ball clay				110		300	180	150	120	100
Bentonite (Wyoming)				60						
Dense grog 8-14 m				350						450
Dense grog 14-35 m				150						50
Dense grog 35-100 m				200						200
Dense grog -100 m				40						
Barium carbonate	10		20		10					
Water	1500	1500	1500	150	1200	1200	1200	1500	1500	150
Soak, hrs	12	12	12	12	12	12	12	12	12	12
Blunge, hrs				1/2	1/2	1/2	1/2			
Mill hrs one gal ball mill (1/2 full of 1 in. pebbles)	15	15	15					15	15	
Screen mesh	100	100	100		100	100	100	100	100	
Filter		✓	✓				✓	✓	✓	
Dry out on plastic bat	✓			✓	✓	✓				✓
Wedge hrs	1/2			1/2	1/2	1/2				1/2
Deflocculent, cc of solution[3]	--	8.5	8.5				8.5	11.0	8.0	
Specific gravity of slip	--	1.72	1.75				1.78	1.72-1.98	1.72	
Plasticity	good	fair	poor	excellent	excellent	excellent	good	fair	fair	excellent
Drying shrinkage linear[5] %,cast	--	4	5				4	4	4	
Drying shrinkage linear[5] %, pressed	6			2	5	5	5			2
Firing shrinkage linear %[6]	4	9	10	4	9	8	9	11	11	4
Fired absorption, %	9	2	0	11	6	5	5	0	0	9
Fired color	dark red	cream	white	dark red[4]	pink[4]	gray	white	white	white	cream
Translucency	none	high	high	none	none	none	none	high	high	none

[1] Reported by C. J. Koenig, Eng. Exp. Stat., O.S.U. Bull. No. 112.
[2] Biscuit fire at 1000°C.
[3] The deflocculent solution is made up by mixing 25 cc of "N" brand sodium silicate and 8 gm of sodium carbonate (anhydrous) and water to make up 100 cc.
[4] Buff with some brick clays.
[5] Based on mold size. (These values vary somewhat with size and shape of specimen.)
[6] Based on dry size. (These values vary somewhat with size and shape of specimen.)

Table A-4

Glaze Compositions

Glaze number	1	2	3	4	--	--	5	6	7	8	9	--
Glaze type	Raw lead, bright	Raw lead, opaque	Raw lead, mat	Fritted[3], bright	Lead frit	Leadless frit	Porcelain, bright	Porcelain, mat	Porcelain, bright	Porcelain, mat	Crystalline	Crystalline frit
Maturing temperature		950-1050°C		1145-1165°C[2]	Fuse	Fuse	1225-1250°C		1450°C		Spec.	Fuse
Cone		08-04		3-4				8-9		16		
White lead	49	48			52							
Red lead			52									
Whiting	10	7	9	7	6	14	18	23	10	6	4	
Soda feldspar	18		28	19		10						
Potash feldspar							43	55	23	14		
Edgar plastic kaolin	5	6	5	6			8	9	7	4	15	
Calcined kaolin	3	5	4				9	7				
Potter's flint	15	20	2	6	34	33	22		47	50	5	
Tin oxide		12										40
Zinc oxide		2		3								25
Titania												9
Aluminum hydrate					8			6	13	26		
Potassium carbonate (anhydrous)						4						18
Sodium carbonate (anhydrous)												3
Borax						2						
Boric oxide (hydrated)						37						
Barium carbonate												5
Lead frit				35								
Leadless frit				24								
Crystalline frit											76	
For body number	1,2	1,2,5	1,2,4,5	2,5,6,7			8	8	9	9	7,8	
Mill, qt jars with 100 cc water for hrs	1/4	1/4	1/4	1	4[1]	4[1]	1	1	1	1	4	2[1]
Specific gravity of slip	1.4	1.5	1.5	1.5			1.5	1.5	1.5	1.5	1.4	
Amount of app. in cc/sq cm.	0.1	0.2	0.2	0.1			0.1	0.2	0.1	0.2	0.3	

[1] Hand grind through 28-mesh screen before milling.
[2] This glaze has a long maturing range and may be used on body #1 at 1020°C.
[3] Suggested by Koenig.

slip may be treated before filtering, but this is not really necessary in small batches. Bodies 5 and 6 are excellent for throwing, and 1 and 7 are fairly satisfactory. Bodies 3, 8, and 9 have such low plasticity that they should be formed by casting. Body 4 contains considerable grog and can therefore be used in large shapes like sculptural pieces. Body 10 is excellent for kiln parts and furniture. It should be realized that many clays contain considerable water when they are received. Allowances must be made for this water content, since the compositions given in Table A–3 are based on dry materials.

Glazes. In Table A–4 is shown a series of glazes for use with the bodies of Table A–3. These are only a few of the thousands of glazes available, but they should serve as a start to the student in this fascinating field.

Table A-5

Colors Developed in Parian Body, No. 3

Stain	Per cent added	Color	Munsel number
Cobaltic oxide	1.0	Deep blue–purple	2.5 PB 3/4
	0.25	Medium blue	2.5 PB 7/2
	0.06	Light blue	2.5 PB 8/2
Chromic oxide	2.0	Sage green	GYG 5/2
	0.5	Gray-green	GYG 6/2
	0.13	Pale gray-green	GYG 7/1
Ferric oxide	10	Deep red-brown	RYR 3/4
	5	Red-brown	RYR 5/2
	1.3	Pale brown	RYR 8/1
Nickel oxide	5	Gray-brown	2.5 Y 4/1
	1.3	Light brown	2.5 Y 7/2
Stain Y	20	Orange	7.5 YR 7/6
	5	Pale peach	7.5 YR 8/6
	1.3	Cream	7.5 YR 9/6
Stain P	10	Pink	5 R 7/4
	2.5	Pale pink	5 R 8/4

Table A-6

Colors Developed in a Porcelain Glaze

Oxide	Amount per cent	Color	Munsel number
Cobalt oxide	4.0	Deep purple-blue	7.5 PB 3/8
	1.5	Purple-blue	7.5 PB 4/8
	0.2	Pale blue	7.5 PB 7/8
Chromic oxide	5.0	Deep green	5 G 5/4
	1.5	Green	2.5 G 5/4
	0.2	Pale green	7.5 GY 8/2
Nickel oxide	4.0	Olive brown	5 Y 6/4
Ferric oxide	4.0	Coral red	RYR 3/8

Table A-7

Colors Developed in Raw Lead Glaze, No. 1

Oxide	Amount per cent	Color	Munsel number
Chromic oxide	0.05	Deep yellow	2.5 Y 8/12
	0.01	Bright yellow	7.5 Y 6/10
Cupric oxide	6	Grass green	2.5 G 6/6
	2	Light green	2.5 G 8/4
Manganese oxide	0.7	Lilac	RPR 5/2
	0.1	Pale lilac	RPR 8/2
Ferric oxide	7	Orange*	YR 5/10
	2	Buff*	YRY 7/6
Nickel oxide	16	Olive green*	5 GY 6/4
	1	Cream	2.5 Y 8/6
Cobalt oxide	0.3	Medium blue	5 PB 4/4
	0.1	Pale blue	5 PB 6/4

*Saturated with crystals.

Colors

The Parian body, No. 3, may be colored by the addition of body stains as shown in Table A-5. The other white firing bodies may be colored in a similar way, but will not be as brilliant. Combinations of the stains may be used to obtain other color blends. The stain is added before milling.

Glaze No. 5 may be colored by the addition of oxides before milling, as shown in Table A-6. Brilliant colors may be produced in glaze No. 1 by the addition of oxides before milling, as shown in Table A-7. Somewhat the same colors will be found in glazes Nos. 2 and 3.

The crystalline glaze must be fired on a careful schedule, as shown on page 182. The colors produced by adding oxides to the frit batch are shown in Table A-8.

Table A-8

Colors in Crystalline Glaze

0.50% CuO gives green and pink crystals

0.25% CoO gives blue crystals

0.50% Fe_2O_3 gives gray crystals

0.50% MnO_2 gives yellow crystals

0.50% Cr_2O_3 gives yellow-green crystals

Stains

In Table A-9 are shown a few stains to be used in bodies or in underglaze or overglaze decoration. The materials should be dry ground thoroughly, calcined at the required temperature, ground very fine (1–5 microns) in a wet ball mill, and carefully washed. Reference should be made to Hainback or Binns for other stains.

Table A-9

Stains

Materials	Red	Pink	Yellow[1]	Green (Victoria)	Blue (Sèvres)	Brown	Black
Borax	4	9					
Whiting	25	1		20			
Alumina, precipitated		81	18		49	57	
Manganese carbonate		9					
Stannic oxide	50						
Potter's flint	18			20			
Fluorspar				12			
Cobalt oxide					37		31
Ferric oxide						22	37
Chromic oxide			2			21	7
Boric oxide (hydrated)					14		
Potassium bichromate	3			35			
Calcium chloride				13			
Manganese dioxide							12
Nickel oxide							13
Titania			73				
Antimony pentoxide			7				
Calcine (hold one hr) at	1230°C	1230°C	1300°C	1230°C	1150°C	1300°C	1300°C
Used in	overglaze	body	body	over- or underglaze	underglaze	underglaze	underglaze
Stable to	1000°C	1275°C	1400°C	1200°C	1200°C	1300°C	1300°C

[1]U. S. Patent 1945809.

Table A-10

Constituent	Per cent
Potter's flint	16
Red lead	61
Boric acid	23

Fluxes (for overglaze colors)

A useful flux is given in Table A–10. The batches should be melted, cooled, and ground dry.

Reference tables

The following tables will be found useful by the ceramist.

Material	Formula	Formula weight	Equivalent weight		
			RO R_2O	R_2O_3	RO_2
Alumina	Al_2O_3	101.9		101.9	
Ammonium carbonate	$(NH_4)_2 \cdot CO_3 \cdot H_2O$	114.1	114.1		
Antimony oxide	Sb_2O_3	291.5		291.5	
Arsenious oxide	As_2O_3	197.8		197.8	
Barium carbonate	$BaCO_3$	197.4	197.4		
Boracic acid	$B_2O_3 \cdot 3H_2O$	123.7		123.7	
Boric oxide	B_2O_3	69.6		69.6	
Borax	$Na_2B_4O_7 \cdot 10 H_2O$	381.4	381.4	190.7	
Calcium carbonate (whiting)	$CaCO_3$	100.1	100.1		
Calcium fluoride	CaF_2	78.1	78.1		
Chromic oxide	Cr_2O_3	152.0	76.0	152.0	
Clay (kaolinite)	$Al_2O_3 \cdot 2SiO_2 \cdot 2H_2O$	258.2		258.2	129.1
Cobaltic oxide	Co_2O_3	165.9	83.0	165.9	
Cryolite	Na_3AlF_6	210.0	140.0	420.0	
Copper oxide (cupric)	CuO	79.6	79.6		
Feldspar (potash)	$K_2O \cdot Al_2O_3 \cdot 6SiO_2$	556.8	556.8	556.8	92.9
Feldspar (soda)	$Na_2O \cdot Al_2O_3 \cdot 6SiO_2$	524.5	524.5	524.5	87.6
Flint (quartz)	SiO_2	60.1			60.1
Iron oxide (ferrous)	FeO	71.8	71.8		
Iron oxide (ferric)	Fe_2O_3	159.7	79.8	159.7	
Lead carbonate (white lead)	$2PbCO_3 \cdot Pb(OH)_2$	775.6	258.5		
Lead oxide (red lead)	Pb_3O_4	685.6	228.5		
Lithium carbonate	Li_2CO_3	73.9	73.9		
Magnesium carbonate	$MgCO_3$	84.3	84.3		
Magnesium oxide	MgO	40.3	40.3		
Manganese dioxide	MnO_2	86.9	86.9		86.9
Nickel oxide	NiO	74.7	74.7		
Potassium carbonate	K_2CO_3	138.0	138.0		
Sodium carbonate	Na_2CO_3	106.0	106.0		
Strontium carbonate	$SrCO_3$	147.6	147.6		
Tin oxide	SnO_2	150.7			150.7
Titanium dioxide	TiO_2	80.1			80.1
Zinc carbonate	$ZnCO_3$	125.4	125.4		
Zinc oxide	ZnO	81.4	81.4		
Zirconium oxide	ZrO_2	123.0			123.0

Table A-12
Atomic Weights and Ionic Radii

Element	Atomic number	Atomic weight*	Atomic radius†	Ionic radius‡	Ion
Actinium	89	227			
Albamine (?)	85	221			
Aluminum	13	26.97	1.43	0.57	Al^{3+}
Antimony	51	121.76	1.45	0.90	Sb^{3+}
Argon	18	39.944	1.91		
Arsenic	33	79.91	1.25	{ 0.69 { ca. 0.4	As^{3+} As^{5+}
Barium	56	137.36	2.17	1.43	Ba^{2+}
Beryllium	4	9.02	1.12	0.34	Be^{2+}
Bismuth	83	209.00	1.55		
Boron	5	10.82	0.97	0.20	B^{3+}
Bromine	35	79.916	1.19	1.96	Br^-
Cadmium	48	112.41	1.49	1.03	Cd^{2+}
Calcium	20	40.08	1.96	1.06	Ca^{2+}
Carbon	6	12.01	0.77	0.2	C^{4+}
Cerium	58	140.13	1.82	1.18	Ce^{3+}
Cesium	55	132.91	2.62	1.65	Cs^+
Chlorine	17	35.457	1.07	1.81	Cl^-
Chromium	24	52.01	1.25	{ 0.64 { 0.3–0.4	Cr^{3+} Cr^{6+}
Cobalt	27	58.94	1.25	0.82	Co^{2+}
Columbium	41	92.91	1.43	{ 0.69 { 0.69	Cb^{4+} Cb^{5+}
Copper	29	63.57	1.28	0.96	Cu^+
Dysprosium	66	162.46		1.07	Dy^{3+}
Erbium	68	167.2	1.86	1.04	Er^{3+}
Europium	63	152.0		1.13	Eu^{3+}
Fluorine	9	19.000		1.33	F^-
Gadolinium	64	156.9		1.11	Gd^{3+}
Gallium	31	69.72	1.22	0.62	Ga^{3+}
Germanium	32	72.60	1.22	0.44	Ge^{4+}
Gold	79	197.2	1.44	1.37	Au^+
Hafnium	72	178.6	1.58	0.84	Hf^{4+}
Helium	2	4.003			
Holmium	67	163.5		1.05	Ho^{3+}
Hydrogen	1	1.0080	0.46	1.54	H^-
Illinium (?)	61	146			
Indium	49	114.76	1.62	0.92	In^{3+}
Iodine	53	126.92	1.36	{ 2.20 { 0.94	I^- I^{5+}
Iridium	77	193.1	1.35	0.66	Ir^{4+}
Iron	26	55.84	1.24	{ 0.83 { 0.67	Fe^{2+} Fe^{2+}
Krypton	36	83.7	2.01		
Lanthanum	57	138.92	1.86	1.22	La^{3+}
Lead	82	207.21	1.75	{ 1.32 { 0.84 { 2.15	Pb^{2+} Pb^{4+} Pb^{4-}
Lithium	3	6.490	1.52	0.78	Li^+
Lutecium	71	174.99		0.99	Lu^{3+}
Magnesium	12	24.32	1.60	0.78	Mg^{2+}
Manganese	25	54.93	1.18	{ 0.91 { 0.70 { 0.52	Mn^{2+} Mn^{3+} Mn^{4+}
Masurium	43				
Mercury	80	200.61	1.50	1.12	Hg^{2+}
Molybdenum	42	95.95	1.36	0.68	Mo^{4+}
Neodymium	60	144.27	1.80	1.15	Nd^{3+}
Neon	10	20.183	1.60		
Nickel	28	58.69	1.24	0.78	Ni^{2+}
Nitrogen	7	14.008	0.71	0.1–0.2	N^{5+}
Osmium	76	190.2	1.35	0.67	Os^{4+}
Oxygen	8	16.000	0.60	1.32	O^{2-}
Palladium	46	106.7	1.37		
Phosphorus	15	30.98		0.3–0.4	P^{5+}
Platinum	78	195.23	1.38		
Polonium	84	210			
Potassium	19	39.096	2.31	1.33	K^+
Praseodymium	59	140.92	1.81	{ 1.16 { 1.00	Pr^{3+} Pr^{4+}
Protoactinium	91	231			

Table A-12 (Continued)

Element	Atomic number	Atomic weight*	Atomic radius†	Ionic radius‡	Ion
Radium	88	226.05			
Radon	86	222			
Rhenium	75	186.31	1.34	0.68	Rh^{3+}
Rhodium	45	102.91			
Rubidium	37	85.48	2.43	1.49	Rb^+
Ruthenium	44	101.7	1.33	0.65	Ru^{4+}
Samarium	62	150.43		1.13	Sm^{3+}
Scandium	21	45.10	1.51	0.83	Sc^{3+}
Selenium	34	78.96	1.16	0.3–0.4	Se^{6+}
				1.91	Se^{2-}
Silicon	14	28.06	1.17	0.39	Si^{4+}
				1.98	Si^{4-}
Silver	47	107.880	1.44	1.13	Ag^+
Sodium	11	22.997	1.86	0.98	Na^+
Strontium	38	87.63	2.15	1.27	Sr^{2+}
Sulfur	16	32.06	1.04	0.34	S^{6+}
				1.74	S^{2-}
Tantalum	73	108.88	1.43	0.68	Ta^{5+}
Tellurium	52	127.61	1.43	0.89	Te^{4+}
				2.11	Te^{2-}
Terbium	65	159.2		1.09	Tb^{3+}
				0.89	Tb^{4+}
Thallium	81	204.39	1.70	1.49	Tl^+
				1.05	Tl^{3+}
Thorium	90	232.12	1.80	1.10	Th^{4+}
Thulium	69	169.4		1.04	Tm^{3+}
Tin	50	118.70	1.40	0.74	Sn^{4+}
				2.15	Sn^{4-}
Titanium	22	47.90	1.46	0.69	Ti^{3+}
				0.64	Ti^{4+}
Tungsten	74	183.92	1.36	0.68	W^{4+}
Uranium	92	238.07	1.38	1.05	U^{4+}
Vanadium	23	50.95	1.30	0.65	V^{3+}
				0.61	V^{4+}
				ca. 0.4	V^{5+}
Virginium	87	224			
Xenon	54	131.3	2.20		
Ytterbium	70	173.04		1.00	Yb^{3+}
Yttrium	39	88.92	1.81	1.06	Y^{3+}
Zinc	30	65.38	1.33	0.83	Zn^{2+}
Zirconium	40	91.22	1.56	0.87	Zr^{4+}

* International atomic weights, 1940, "Handbook of Chemistry and Physics," 24th ed., Chemical Rubber Publishing Co., Cleveland, 1940.

† One-half distance of closest approach in structure of the element; Evans[103]; in angstrom units, 10^{-8} cm.

‡ Ionic radii for sixfold coordination; Evans[103]; in angstrom units, 10^{-8} cm.

Table A-13
Temperature Equivalents of Orton Pyrometric Cones

Cone No.	End point, 20°C. per hr.		End point, 150°C. per hr.	
	°C.	°F.	°C.	°F.
022	585	1090	605	1120
021	595	1100	615	1140
020	625	1160	650	1200
019	630	1170	660	1220
018	670	1240	720	1330
017	720	1330	770	1420
016	735	1360	795	1460
015	770	1420	805	1480
014	795	1460	830	1530
013	825	1520	860	1580
012	840	1540	875	1610
011	875	1610	905	1660
010	890	1630	895	1640
09	930	1710	930	1710
08	945	1730	950	1740
07	975	1790	990	1810
06	1005	1840	1015	1860
05	1030	1890	1040	1900
04	1050	1920	1060	1940
03	1080	1980	1115	2040
02	1095	2000	1125	2060
01	1110	2030	1145	2090
1	1125	2060	1160	2120
2	1135	2080	1165	2130
3	1145	2090	1170	2140
4	1165	2130	1190	2170
5	1180	2160	1205	2200
6	1190	2170	1230	2250
7	1210	2210	1250	2280
8	1225	2240	1260	2300
9	1250	2280	1285	2350
10	1260	2300	1305	2380
11	1285	2350	1325	2420
12	1310	2390	1335	2440
13	1350	2460	1350	2460
14	1390	2530	1400	2550
15	1410	2570	1435	2620
16	1450	2640	1465	2670
17	1465	2670	1475	2690
18	1485	2710	1490	2710
19	1515	2760	1520	2770
20	1520	2770	1530	2790
23			1580*	2880*
26			1595	2900
27			1605	2920
28			1615	2940
29			1640	2980
30			1650	3000
31			1680	3060
32			1700	3090
33			1745	3170
34	1755	3190	1760	3200
35	1775	3230	1785	3250
36	1810	3290	1810	3290
37	1830	3330	1820	3310
38	1850	3360	1835	3340
39	1865	3390		
40	1885	3430		
41	1970	3580		
42	2015	3660		

*Cones 23 to 38 heated at 100°C. per hr.

Table A-14
Temperature-Conversion Table
(Dr. L. Waldo, in *Metallurgical and Chemical Engineering*, March, 1910)

C	0	10	20	30	40	50	60	70	80	90
	F	F	F	F	F	F	F	F	F	F
−200	−328	−346	−364	−382	−400	−418	−436	−454	....	....
−100	−148	−166	−184	−202	−220	−238	−256	−274	−292	−310
− 0	+ 32	+ 14	− 4	− 22	− 40	− 58	− 76	− 94	−112	−130
0	32	50	68	86	104	122	140	158	176	194
100	212	230	248	266	284	302	320	338	356	374
200	392	410	428	446	464	482	500	518	536	554
300	572	590	608	626	644	662	680	698	716	734
400	752	770	788	806	824	842	860	878	896	914
500	932	950	968	986	1004	1022	1040	1058	1076	1094
600	1112	1130	1148	1166	1184	1202	1220	1238	1256	1274
700	1292	1310	1328	1346	1364	1382	1400	1418	1436	1454
800	1472	1490	1508	1526	1544	1562	1580	1598	1616	1634
900	1652	1670	1688	1706	1724	1742	1760	1778	1796	1814
1000	1832	1850	1868	1886	1904	1922	1940	1958	1976	1994
1100	2012	2030	2048	2066	2084	2102	2120	2138	2156	2174
1200	2192	2210	2228	2246	2264	2282	2300	2318	2336	2354
1300	2372	2390	2408	2426	2444	2462	2480	2498	2516	2534
1400	2552	2570	2588	2606	2624	2642	2660	2678	2696	2714
1500	2732	2750	2768	2786	2804	2822	2840	2858	2876	2894
1600	2912	2930	2948	2966	2984	3002	3020	3038	3056	3074
1700	3092	3110	3128	3146	3164	3182	3200	3218	3236	3254
1800	3272	3290	3308	3326	3344	3362	3380	3398	3416	3434
1900	3452	3470	3488	3506	3524	3542	3560	3578	3596	3614
2000	3632	3650	3668	3686	3704	3722	3740	3758	3776	3794
2100	3812	3830	3848	3866	3884	3902	3920	3938	3956	3974
2200	3992	4010	4028	4046	4064	4082	4100	4118	4136	4154
2300	4172	4190	4208	4226	4244	4262	4280	4298	4316	4334
2400	4352	4370	4388	4406	4424	4442	4460	4478	4496	4514
2500	4532	4550	4568	4586	4604	4622	4640	4658	4676	4694
2600	4712	4730	4748	4766	4784	4802	4820	4838	4856	4874
2700	4892	4910	4928	4946	4964	4982	5000	5018	5036	5054
2800	5072	5090	5108	5126	5144	5162	5180	5198	5216	5234
2900	5252	5270	5288	5306	5324	5342	5360	5378	5396	5414
3000	5432	5450	5468	5486	5504	5522	5540	5558	5576	5594
3100	5612	5630	5648	5666	5684	5702	5720	5738	5756	5774
3200	5792	5810	5828	5846	5864	5882	5900	5918	5936	5954
3300	5972	5990	6008	6026	6044	6062	6080	6098	6116	6134
3400	6152	6170	6188	6206	6224	6242	6260	6278	6296	6314
3500	6332	6350	6368	6386	6404	6422	6440	6458	6476	6494
3600	6512	6530	6548	6566	6584	6602	6620	6638	6656	6674
3700	6692	6710	6728	6646	6764	6782	6800	6818	6836	6854
3800	6872	6890	6908	6926	6944	6962	6980	6998	7016	7034
3900	7052	7070	7088	7106	7124	7142	7160	7178	7196	7214

°C.	°F.
1	1.8
2	3.6
3	5.4
4	7.2
5	9.0
6	10.8
7	12.6
8	14.4
9	16.2
10	18.0

°F.	°C.
1	0.56
2	1.11
3	1.67
4	2.22
5	2.78
6	3.33
7	3.89
8	4.44
9	5.00
10	5.56
11	6.11
12	6.67
13	7.22
14	7.78
15	8.33
16	8.89
17	9.44
18	10.00

Examples. 1347°C. = 2444°F. + 12.6°F. = 2456.6°F.; 3367°F. = 1850°C + 2.78°C = 1852.78°C.

Table A-15
Standard Calibration Data for Chromel-Alumel Couples

E.m.f., millivolts	Reference junction at 0°C.					
	0	10	20	30	40	50
	Temperature, 0°C.					
0	0	246	485	720	966	1232
0.2	5	251	490	725	972	1237
.4	10	256	494	730	977	1243
.6	15	261	499	735	982	1249
.8	20	266	504	740	987	1254
1.0	25	271	508	744	992	1260
1.2	30	276	513	749	997	1266
1.4	35	280	518	754	1002	1271
1.6	40	285	523	759	1007	1277
1.8	45	290	527	764	1013	1283
2.0	50	295	532	768	1018	1288
2.2	54	300	537	773	1023	1294
2.4	59	305	541	778	1028	1300
2.6	64	310	546	783	1033	1306
2.8	69	315	551	788	1038	1311
3.0	74	319	555	792	1044	1317
3.2	79	324	560	797	1049	1323
3.4	83	329	565	802	1054	1329
3.6	88	334	570	807	1059	1334
3.8	93	338	574	812	1065	1340
4.0	98	343	579	817	1070	1346
4.2	102	348	584	822	1075	1352
4.4	107	353	588	827	1081	1358
4.6	112	358	593	832	1086	1364
4.8	117	362	598	837	1091	1370
5.0	122	367	602	841	1096	1376
5.2	127	372	607	846	1102	1382
5.4	132	376	612	851	1107	1388
5.6	137	381	616	856	1112	1394
5.8	142	386	621	861	1118	1400
6.0	147	391	626	866	1123	
6.2	152	396	631	871	1128	
6.4	157	400	635	876	1134	
6.6	162	405	640	881	1139	
6.8	167	410	645	886	1144	
7.0	172	414	649	891	1150	
7.2	177	419	654	896	1155	
7.4	182	424	659	901	1161	
7.6	187	429	664	906	1166	
7.8	192	433	668	911	1171	
8.0	197	438	673	916	1177	
8.2	202	443	678	921	1182	
8.4	207	448	683	926	1188	
8.6	212	452	687	931	1193	
8.8	217	457	692	936	1199	
9.0	222	462	697	941	1204	
9.2	227	466	701	946	1210	
9.4	232	471	706	951	1215	
9.6	237	476	711	956	1221	
9.8	241	480	716	961	1226	
10.0	246	485	720	966	1232	

By permission from *Refractories*, 3rd Ed., by F. H. Norton. Copyright, 1949, McGraw-Hill Book Company, Inc.

Table A-16
Standard Calibration Data for Chromel-Alumel Couples

E.m.f., millivolts	Reference junction at 32°F.					
	0	10	20	30	40	50
	Temperature, °F.					
0	32	475	905	1329	1772	2250
0.2	41	484	913	1338	1781	2260
0.4	50	493	922	1346	1790	2270
0.6	59	502	930	1355	1799	2280
0.8	68	510	939	1363	1808	2290
1.0	77	519	947	1372	1818	2300
1.2	86	528	956	1380	1827	2310
1.4	95	537	964	1389	1836	2320
1.6	104	546	973	1398	1845	2331
1.8	113	554	981	1407	1855	2341
2.0	121	563	990	1415	1864	2351
2.2	130	572	998	1424	1873	2362
2.4	139	580	1006	1433	1882	2372
2.6	147	589	1015	1441	1892	2382
2.8	156	598	1023	1450	1901	2393
3.0	165	607	1032	1459	1911	2403
3.2	173	615	1040	1467	1920	2413
3.4	182	624	1049	1476	1930	2424
3.6	190	632	1057	1485	1939	2434
3.8	199	641	1065	1494	1949	2445
4.0	208	650	1074	1503	1958	2455
4.2	217	658	1083	1511	1967	2466
4.4	225	667	1091	1520	1977	2476
4.6	234	675	1099	1529	1986	2487
4.8	243	684	1108	1538	1996	2497
5.0	251	693	1116	1547	2005	
5.2	260	701	1125	1555	2015	
5.4	269	710	1133	1564	2024	
5.6	278	718	1142	1573	2034	
5.8	287	727	1150	1582	2044	
6.0	296	735	1158	1591	2053	
6.2	305	744	1167	1600	2063	
6.4	314	752	1175	1609	2072	
6.6	323	760	1184	1618	2082	
6.8	332	769	1193	1627	2092	
7.0	341	778	1201	1636	2101	
7.2	350	786	1210	1645	2111	
7.4	359	795	1218	1654	2121	
7.6	368	803	1227	1663	2130	
7.8	377	812	1235	1672	2140	
8.0	386	820	1243	1680	2150	
8.2	395	829	1252	1689	2160	
8.4	404	838	1260	1698	2170	
8.6	413	846	1269	1708	2180	
8.8	422	855	1278	1717	2190	
9.0	431	863	1286	1726	2200	
9.2	440	872	1295	1735	2210	
9.4	449	880	1303	1744	2220	
9.6	457	889	1312	1753	2230	
9.8	466	897	1320	1762	2240	
10.0	475	905	1329	1772	2250	

By permission from *Refractories*, 3rd Ed., by F. H. Norton. Copyright, 1949, McGraw-Hill Book Company, Inc.

Table A-17

Standard Calibration Data for Thermocouples from Platinum
and Platinum Alloyed with 10 Per Cent Rhodium

E.m.f., microvolts	0	1,000	2,000	3,000	4,000	5,000	6,000	7,000	8,000	**9,000**
					Temperatures °C.					
0	0.0	147.1	265.4	374.3	478.1	578.3	675.3	769.5	861.1	950.4
100	17.8	159.7	276.6	384.9	488.3	588.1	684.8	778.8	870.1	959.2
200	34.5	172.1	287.7	395.4	498.4	597.9	694.3	788.0	879.1	968.0
300	50.3	184.3	298.7	405.9	508.5	607.7	703.8	797.2	888.1	976.7
400	65.4	196.3	309.7	416.3	518.6	617.4	713.3	806.4	897.1	985.4
500	80.0	208.1	320.6	426.7	528.6	627.1	722.7	815.6	906.1	994.1
600	94.1	219.7	331.5	437.1	538.6	636.8	732.1	824.7	915.0	1002.8
700	107.8	231.2	342.3	447.4	548.6	646.5	741.5	833.8	923.9	1011.5
800	121.2	242.7	353.0	457.7	558.5	656.1	750.9	842.9	932.8	1020.1
900	134.3	254.1	363.7	467.9	568.4	665.7	760.2	852.0	941.6	1028.7
1,000	147.1	265.4	374.3	478.1	578.3	675.3	769.5	861.1	950.4	1037.3

E.m.f., microvolts	10,000	11,000	12,000	13,000	14,000	15,000	16,000	17,000	**18,000**
					Temperatures °C.				
0	1037.3	1122.2	1205.9	1289.3	1372.4	1454.8	1537.5	1620.9	1704.3
100	1045.9	1130.6	1214.2	1297.7	1380.7	1463.0	1545.8	1629.2	1712.6
200	1054.4	1139.0	1222.6	1306.0	1389.0	1471.2	1554.1	1637.6	1721.0
300	1062.9	1147.4	1230.9	1314.3	1397.3	1479.4	1562.4	1645.9	1729.3
400	1071.4	1155.8	1239.3	1322.6	1405.6	1487.7	1570.8	1654.3	1737.7
500	1079.9	1164.2	1247.6	1330.9	1413.8	1496.0	1579.1	1662.6	1746.0
600	1088.4	1172.5	1255.9	1339.2	1422.0	1504.3	1587.5	1670.9	1754.3
700	1096.9	1180.9	1264.3	1347.5	1430.2	1512.6	1595.8	1679.3	
800	1105.4	1189.2	1272.6	1355.8	1438.4	1520.9	1604.2	1687.6	
900	1113.8	1197.6	1281.0	1364.1	1446.6	1529.2	1612.5	1696.0	
1,000	1122.2	1205.9	1289.3	1372.4	1454.8	1537.5	1620.9	1704.3	

Table A-18

Standard Calibration Data for Thermocouples from Platinum
and Platinum Alloyed with 10 Per Cent Rhodium

E.m.f., microvolts	0	1,000	2,000	3,000	4,000	5,000	6,000	7,000	8,000	9,000
					Temperatures °F.					
0	32.0	296.8	509.7	705.7	892.6	1072.9	1247.5	1417.1	1582.0	1742.7
100	42.0	319.5	529.9	724.8	910.9	1090.6	1264.6	1433.8	1598.2	1758.6
200	94.1	341.8	549.9	743.7	929.1	1108.2	1281.7	1450.4	1614.4	1774.4
300	122.5	363.7	569.7	762.6	947.3	1125.9	1298.8	1467.0	1630.6	1790.1
400	149.7	385.3	589.5	781.3	965.5	1143.3	1315.9	1483.5	1646.8	1805.7
500	176.0	406.6	609.1	800.1	983.5	1160.8	1332.9	1500.1	1663.0	1821.4
600	201.4	427.5	628.7	818.8	1001.5	1178.2	1349.8	1516.5	1679.0	1837.0
700	226.0	448.2	648.1	837.3	1019.5	1195.7	1366.7	1532.8	1694.8	1852.7
800	250.2	468.9	667.4	855.9	1037.3	1213.0	1383.6	1549.2	1711.0	1868.2
900	273.7	489.4	686.7	874.2	1055.1	1230.3	1400.4	1565.6	1726.9	1883.7
1,000	296.8	509.7	705.7	892.6	1072.9	1247.5	1417.1	1582.0	1742.7	1899.1

E.m.f., microvolts	10,000	11,000	12,000	13,000	14,000	15,000	16,000	17,000	18,000
					Temperatures °F.				
0	1899.1	2052.0	2202.6	2352.7	2502.3	2650.6	2799.5	2949.6	3099.7
100	1914.6	2067.1	2217.6	2367.9	2517.3	2665.4	2814.4	2964.6	3114.7
200	1929.9	2082.2	2232.7	2382.8	2532.2	2680.2	2829.4	2979.7	3129.8
300	1945.2	2097.3	2247.6	2397.7	2547.1	2694.9	2844.3	2994.6	3144.7
400	1960.5	2112.4	2262.7	2412.7	2562.1	2709.9	2859.4	3009.7	3159.9
500	1975.8	2127.6	2277.7	2427.6	2576.8	2724.8	2874.4	3024.7	3174.8
600	1991.1	2142.5	2292.6	2442.6	2591.6	2739.7	2889.5	3039.6	3189.7
700	2006.4	2157.6	2307.7	2457.5	2606.4	2754.7	2904.4	3054.7	
800	2021.7	2172.6	2322.7	2472.4	2621.1	2769.6	2919.6	3069.7	
900	2036.8	2187.7	2337.8	2487.4	2635.9	2784.6	2934.5	3084.8	
1,000	2052.0	2202.6	2352.7	2502.3	2650.6	2799.5	2949.6	3099.7	

By permission from *Refractories*, 3rd Ed., by F. H. Norton.   Copyright, 1949, McGraw-Hill Book Company, Inc.

Table A-19

Standard Calibration Data for Copper-Constantan Thermocouple

E.m.f., microvolts	0	1,000	2,000	3,000	4,000	5,000	6,000	7,000	8,000	9,000
	Temperatures °C.									
0	0.0	25.3	49.2	72.1	94.1	115.3	135.9	155.9	175.5	194.6
100	2.6	27.7	51.5	74.3	96.2	117.4	137.9	157.0	177.4	196.5
200	5.2	30.2	53.9	76.5	98.4	119.5	140.0	159.9	179.4	198.4
300	7.7	32.6	56.2	78.8	100.5	121.6	142.0	161.9	181.3	200.3
400	10.3	35.0	58.5	81.0	102.7	123.6	144.0	163.8	183.2	202.2
500	12.8	37.4	60.8	83.2	104.8	125.7	146.0	165.8	185.1	204.0
600	15.3	39.8	63.0	85.4	106.9	127.7	148.0	167.7	187.0	205.9
700	17.8	42.2	65.3	87.6	109.0	129.8	150.0	169.7	188.0	207.8
800	20.3	44.5	67.6	89.7	111.1	131.8	152.0	171.6	190.8	209.8
900	22.8	46.9	69.8	91.9	113.2	133.9	154.0	173.6	192.7	211.5
1,000	25.3	49.2	72.1	94.1	115.3	135.9	155.9	175.5	194.6	213.4

E.m.f., microvolts	10,000	11,000	12,000	13,000	14,000	15,000	16,000	17,000	18,000
	Temperatures °C.								
0	213.4	231.7	249.8	267.6	285.1	302.4	319.5	336.4	353.1
100	215.2	233.6	251.6	269.4	286.9	304.1	321.2	338.0	
200	217.2	235.4	253.4	271.1	288.6	305.9	322.9	339.7	
300	218.9	237.2	255.2	272.9	290.3	307.6	324.6	341.4	
400	220.8	239.0	257.0	274.6	292.1	309.3	326.3	343.1	
500	222.6	240.8	258.7	276.4	293.8	311.0	327.9	344.7	
600	224.4	242.6	260.5	278.2	295.5	312.7	329.6	346.4	
700	226.3	244.4	262.3	279.9	297.3	314.4	331.3	348.1	
800	228.1	246.2	264.1	281.6	299.0	316.1	333.0	349.7	
900	229.9	248.0	265.8	283.4	300.7	317.8	334.7	351.4	
1,000	231.7	249.8	267.6	285.1	302.4	319.5	336.4	353.1	

By permission from *Refractories*, 3rd Ed., by F. H. Norton.   Copyright, 1949, McGraw-Hill Book Company, Inc.

Table A-20

Standard Calibration Data for Copper-Constantan Thermocouple

E.m.f., microvolts	0	1,000	2,000	3,000	4,000	5,000	6,000	7,000	8,000	9,000
	Temperatures °F.									
0	32.0	77.5	120.6	161.7	201.3	239.6	276.6	312.7	347.9	382.3
100	36.7	81.9	124.8	165.7	205.2	243.3	280.3	314.6	351.4	385.7
200	41.4	86.3	128.9	169.8	209.1	247.1	283.9	319.8	354.8	389.1
300	45.9	90.6	133.1	173.9	212.9	250.8	287.6	323.3	358.3	392.5
400	50.5	95.0	137.2	177.7	216.8	254.5	291.2	326.9	361.8	395.9
500	55.0	99.3	141.4	181.7	220.6	258.2	294.8	330.4	365.2	399.3
600	59.5	103.6	145.5	185.7	224.4	261.9	298.4	333.9	368.6	402.6
700	64.0	107.9	149.5	189.6	228.2	265.6	302.0	337.4	370.4	406.0
800	68.5	112.1	153.6	193.5	232.0	269.3	305.6	340.9	375.6	409.3
900	73.0	116.3	157.7	197.4	235.8	273.0	309.1	344.4	378.9	412.7
1,000	77.5	120.6	161.7	201.3	239.6	276.6	312.7	347.9	382.3	416.0

E.m.f., microvolts	10,000	11,000	12,000	13,000	14,000	15,000	16,000	17,000	18,000	19,000
	Temperatures °F.									
0	416.0	449.1	481.7	513.7	545.2	576.4	607.1	637.4	667.6	
100	419.4	452.4	484.9	516.8	548.4	579.4	610.1	640.5		
200	423.0	455.7	488.1	519.0	551.5	582.5	613.2	643.5		
300	426.0	459.0	491.3	523.2	554.6	585.6	616.2	646.5		
400	429.3	462.2	494.5	526.4	557.7	588.7	619.3	649.5		
500	432.7	465.5	497.7	529.5	560.8	591.8	622.3	652.5		
600	436.0	468.7	500.9	532.7	564.0	594.8	625.4	655.5		
700	439.3	472.0	504.1	535.8	567.1	597.9	628.4	658.5		
800	442.6	475.2	507.3	539.0	570.2	601.0	631.4	661.6		
900	445.9	478.4	510.5	542.1	573.6	604.0	634.4	664.6		
1,000	449.1	481.7	513.7	545.2	576.4	607.1	637.4	667.6		

By permission from *Refractories*, 3rd Ed., by F. H. Norton. Copyright, 1949, McGraw-Hill Book Company, Inc.

INDEX

TABLE OF SYMBOLS AND UNITS

Symbol	Concept	Units	Dimensions
l	length	cm (in., ft)	L
Δl	change in length	cm	L
μ (mu)	length	micron $= 10^{-4}$ cm	L
$m\mu$ (mu)	length	millimicron $= 10^{-7}$ cm	L
Å	length	Ångstrom unit $= 10^{-8}$ cm	L
r	radius	cm (in., ft)	L
S	area	cm²	L^2
V	volume	cm³	L^3
ΔV	change in volume	cm³	L^3
W	weight	gm (lb)	M
ρ (rho)	density	gm·cm⁻³	ML^{-3}
d	bulk density	gm·cm⁻³	ML^{-3}
$s.g.$	specific gravity	number	—
P	per cent porosity (based on bulk volume)	number	—
F	shearing force (degrees of freedom in the phase rule)	dynes·cm⁻² number	MLT^{-1} —
F_o	yield point	dynes·cm⁻²	MLT^{-1}
σ (sigma)	normal stress	dynes·cm⁻²	MLT^{-1}
M	elastic modulus	dynes·cm⁻²	MLT^{-1}
G	modulus of rigidity	dynes·cm⁻²	MLT^{-1}
v	velocity	cm·sec⁻¹	LT^{-2}
dl/dt	rate of normal flow	cm·sec⁻¹	LT^{-2}
dv/dr	rate of shear strain	cm·sec⁻¹·cm⁻¹	T^{-1}
t	time	sec	T
R^1	rate of growth	cm·sec⁻¹	LT^{-1}
η (eta)	coef. of viscosity	(poise) dynes·sec·cm⁻²	$ML^{-1}\,T^{-1}$
η_s (eta)	coef. of viscosity of suspension	(poise) dynes·sec·cm⁻²	$ML^{-1}\,T^{-1}$
η_l (eta)	coef. of viscosity of liquid	(poise) dynes·sec·cm⁻²	$ML^{-1}\,T^{-1}$
θ (theta)	coef. of thixotropy	dynes·cm⁻²	$ML^{-2}\,T^{-1}$
C	Wt. concentration of solid in susp. (components in the phase rule)	number	—
C^1	Wt. concentration of water in solid	number	—
T	absolute temperature	°Kelvin	—
K	thermal conductivity	cal·cm⁻², cm·sec⁻¹·°C⁻¹	—
P	phases in the phase rule	number	—
n	slope of line	number	—
A,B,C	components or cations	—	—
R	any cation	—	—
dV/dt	rate of volume flow	cm³·sec⁻¹	L^3T^{-1}
a,b,c	dimensions of unit cell	cm (Å)	L
x,y,z	crystallographic axes	—	—
β (beta)	monoclinic angle	degrees	—
N	index of refraction	number	—
pH	hydrogen ion concentration	number	—